ANNUALS FROM SEED

BY

ROY GENDERS

MAGNA PRINT BOOKS
LITTON, YORKSHIRE

First Published in large print 1974
by
Magna Print Books,
Litton, Skipton, Yorkshire.
by arrangement with
W. & G. Foyle Limited,
London.

ISBN 0 86009 021 3

Printed in Great Britain by
Graphicapress (Bradford) Ltd.,
Bradford.

CONTENTS

Chapter 1

ANNUALS; THEIR VALUE IN THE GARDEN

What is an annual? — Their value in interplanting — The advent of new varieties — Their Adaptability — The value of glass.

THE Annual is a plant that will bloom and have existence during a single year. It is not quite correct to say that this will be the year in which the seed is sown for many annuals; those of a hardy nature, may have been sown the previous year when they have more of a biennial habit. Indeed in this book, the biennial plants are also included for it is difficult to write a book on annual plants with their varying habits and omit the true biennials. Both annuals and biennials are produced entirely from seed and it is more correct to say that annuals bloom within twelve months of sowing the seed.

A number of plants which are really of perennial habit will also be included, plants such as the pansy and antirrhinum, which are raised in vast quantities each year for spring and summer bedding and which are treated almost entirely as annuals or biennials.

Not all biennials are completely hardy, whilst the annuals may be classed as

(a) Hardy
(b) Half-hardy.

The great value of all these plants in the modern garden is that they may be inexpensively produced, whilst they will give the maximum of colour over the longest possible period. Again, many may be sown where they are to bloom, thus saving valuable time in the garden, whilst it is also possible to obtain a mass of colour from a piece of ground which has not, through reasons of lack of time or illness, or adverse weather, been sufficiently cleaned, or brought to the necessary condition to grow perennial plants. this is a most important factor in the use of certain annuals in those gardens attached to a newly-built house, of which there are many at the present time, and which may take several years to bring into a satisfactory state of fertility. Yet that bank of clay and stones may become a mass of vivid colouring from a small packet of nasturtium seed, or the small, almost un-made bed beneath the kitchen window could yield a continuous supply of bloom for decorating the home, with marigolds or coreopsis sown where they are to bloom, the plants being covered with bloom in less than three months from the time the seed is sown. A packet of seed costing 11p and there will be colour both in the home and in the garden from July until November;

surely there is no better value today. Indeed for a pound or two, one could fill the garden with flowers of great beauty, in bloom from April until November. The use of annuals (and biennials) is the most economical of all forms of flower growing, whilst no plants will give so prolific a display of bloom over so long a period. In comparison (amongst perennials), only the long flowering scabious, a plant more suitable for cutting than for garden decoration, can equal the average annual in its long flowering period.

The Value of inter-planting

Another factor of considerable value with annuals is their use in interplanting. There was a time, possibly until the Second World War, when there was no shortage of garden labour. Even during the years of the Industrial Revolution which saw the first drift of the countryman to the towns with the promise of more money and better living conditions, the unsatisfactory conditions of the factory ensured there was always some labour available for our gardens. Depressed conditions after the First World War produced no shortage of gardeners, but now, all this has changed. The amenities of the towns and the better working conditions, the high wages paid in industry, have brought about the first great shortage of garden labour. Owners of gardens, big and small, must of necessity do their own gardening and this makes it imperative to make the garden as labour saving as possible.

This is where those plants suitable for inter-planting are now so useful. the days when spring flowering plants could be lifted, plants such as polyanthus, moved to another part of the garden, the beds being re-planted with plants for summer flowering, are gone for most gardeners. The polyanthus and the daffodils are left where they are to bloom again the following year, annuals being used for planting in between to provide the summer display. This they so conveniently do, for they quickly cover up the dying leaves of bulbs and spring flowering plants, giving them the necessary protection from the summer sun, choking out annual weeds, and covering the beds with rich colour which will be almost a continuation from the spring display.

Again, the days of the conservatory are over for most house-holders, the high cost of building, the need for a garage rather than somewhere to grow plants, and the ever increasing cost of heating, has brought about a considerable reduction in the use of those summer flowering plants which require wintering indoors; the pelargonium and begonia and similar plants.

The annual has come to be recognised as taking the place of these less hardy plants, plants which also require considerably more labour in their use.

The annual is an asset to any garden on account of its versatility, but only during recent times has its value in this way been appreciated. For late sowings, to fill up those bare places about the herbaceous border; for sowing about the shrubbery, planted in ever-increasing numbers on account of their ease and labour saving qualities; and their use for providing a frame or edging to flower beds, has given the annual a new popularity. But it was not until William Robinson, who showed such a distaste for the all too formal gardening which reached its climax towards the end of the 19th century, preached the need for less formality that annuals came to receive any appreciable recognition. As we have seen, there being no problems of labour or of providing winter accommodation, there was little point in deviating from the recognised form of bedding and the use of cut flowers in the home was little practised, the scented-leaf geranium and the aspidistra providing the indoor decoration. All the same, Robinson could see the great value and charm of many annuals and pleaded for their greater use, though this did not come about, and then only because gardening conditions had changed, until half a century after his death. Today more annuals are sold each year than any other garden plant, vegetables included.

Advent of new varieties

It must be said that much of the popularity now being enjoyed by so many annuals is due to the efforts of our large seed houses to raise a plant which will be suitable for garden display. Annuals during the first half of the twentieth century were suitable for cut flower growers, the raisers being well aware of the need to cater for the new interest in flowers for indoor decoration to take the place of the now less popular aspidistra, but there were few annuals with a sufficiently compact habit to make them suitable for bedding and only since the end of the Second World War have annuals been used on a large scale for this purpose. The value of biennials for spring bedding has for long been appreciated, possibly more so than at the present time, the Victorians making great use of wallflowers and forget-me-nots for carpeting a bed of daffodils or tulips.

The value of fragrance in the garden was also much appreciated and mignonette, night scented stock and the evening primrose, were frequently sown beneath a window where their fragrance could be enjoyed during an early summer evening. Lobelia and alyssum were also in vogue for edging those brilliant beds of geraniums and calceolarias, but there were few other annuals for bedding.

The modern gardener has not only a wide range of annuals from which to choose, but a far greater colour range. Today there is a mauve alyssum and a pale blue lobelia as a welcome change from the over-planted white alyssum and dark blue lobelia and there are similar introductions with almost all other annuals. There is a greater choice of form too, the zinnia now being obtainable in the form of the dahlia or the chrysanthemum, with its fluted petals; the aster in the large ostrich plume type or the tiny flowered lilliput form.

Compactness has also been a feature of the modern annual, these tall growing, sparse foliaged plants having been replaced by those having a neat, rather more spreading habit and flowering over a longer season. Here the new Waldersee aster and the dainty, richly coloured French marigolds, in bloom from July until almost the end of November, have earned a new popularity for these plants; old favourites in a new form.

The question of hardiness and disease has also received considerable attention. There is now a new range of antirrhinums which are resistant to rust disease, a trouble which so often reduced the vitality and length of flowering period of the non-resistant varieties; and a race of the Flowering stock much less prone to wilting in its early stages. Again with the Flowering stock, Hansen's new strain makes it possible to eliminate those unwanted single flowering plants which always made a far too frequent appearance in beds of stock until very recent times.

The greatly extended flowering period of the nemesia and the *Phlox drummondii* has brought new popularity to these lovely bedding annuals, whilst the introduction of more compact forms of the godetia and nicotiana, has made these plants extremely popular for bedding whereas until recent years they were planted only about the mixed border. It is now possible to sow only those plants with so dwarf a habit as to make any staking quite unnecessary, a most important consideration in the labour saving garden.

In addition, depth of colour, sturdiness of stem and quality of bloom have all contributed to give the annual plant the great popularity it now enjoys. The introduction of a greater variety of colour with these plants is not generally realised, but there was a time when there was little in the way of art or pastel shades, with the result that the bedding of annuals gave an appearance of harsh brilliance which looked quite out of place in so many gardens. Today, the softer and less formal appearance of so many annuals is greatly appreciated by the discerning gardener.

The adaptability of annuals

For rapidity to come into bloom and for providing a continuation of the spring display, annuals are difficult to better for they will come into bloom early in July, almost as soon as the later flowering tulips and the polyanthus have finished flowering. The plants should be set out between those of the spring display, early in May if the annuals are hardy; the end of the month if only half-hardy; or seed may be sown in the open ground where the plants are to bloom. There will thus be a continuation of bloom from the same beds from early April until early November, seven months of bloom for the outlay of a few pence. True, the herbacious border, carefully planted, will also provide a continuation of colour, but there will be long periods when colour is at a minimum and those plants which have finished blooming will look anything but tidy. Again, no staking will be necessary with almost the entire range of annuals and biennials, whilst the ground may be cleared entirely of plants and weeds, except where being used for inter-planting. Thus the ground is perfectly clean to commence again the following summer, or it may be planted at once with bulbs or spring flowering biennials such as wallflowers. The ground is thus kept free from weeds which are difficllt to eradicate from a herbacious border, whilst the soil does not become sour or ridden with pests or disease. Beginning each season with clean ground will make for very little weeding through summer when there are so many other jobs about the garden, lawn mowing and hedge clipping, to demand attention.

Annuals also fit well into rotational cropping, especially with tulips and spring flowering biennials, which finish flowering during June when the beds may be cleared completely, thoroughly cleaned and replanted the same day with summer annuals.

The same may also be the case when using window boxes, for suitable annuals may be planted amongst the dying foliage of spring flowering bulbs such as crocus and dwarf tulips, hyacinths or primroses, and will be ready to take over the flowering soon after the spring display has ended.

Annuals may be used in numerous ways quite apart from in the garden for window boxes there is a wide range of plants with a dwarf, neat habit, whilst for tubs about the garden or round a courtyard, those plants having a trailing habit are charming when draped around the outside of the tubs with taller subjects for the centre. Here again the plants may be set out as soon as the bulb display has ended, without any disturbance of the soil or bulbs, the soil being changed every third year, or alternate years where tulips are grown, for the bulbs will require renewing after two seasons.

The value of glass

If a cold frame or a greenhouse is available the less hardy annuals may be raised and the plants will be ready to go out the moment the beds are free. a cold frame may be inexpensively made and will ensure that a much wider range of plants will be available, even those annuals considered to be amongst the most tender plants. Making a cold frame is described in Chapter 3.

But even where no frame is available, a far greater use of annuals could well be made by raising plants under cloches and in the open and whilst those of dual-purpose value may be used with great effect both for bedding and cutting, greater use could be made by dividing one's annuals into those to be grown entirely for providing cut bloom and those to be used only for bedding or for garden display. This would mean that a small part of the garden was set off for cut flower purposes, the seed being sown where the plants are to bloom and being wintered, in an exposed garden, under cloches. The same cloches would be used to cover boxes of seed sown late in March of both hardy and half-hardy plants for bedding out in June.

Whether being grown for cutting or bedding display almost all annuals favour a position of full sun, though there are a few exceptions to the rule.

But generally annuals must be given an open, sunny position where no amount of hot sunshine will dim the brilliance of their bloom. In fact there are few plants apart from the geranium which will revel in the hot, dry conditions as do so many of the annuals. For a sun-baked soil there is no more valuable plant, whilst they do not require anything like the amount of humus and manure in the soil as do the perennials, which during the present shortage of humus forming manures and the high cost of artificials is an important consideration. Where the ground is of a poor nature, then annuals must always be grown instead of perennials, there being a number of annuals which will flourish on almost barren rock.

Where one gardens over a chalk sub-soil, generally a hot, dry soil, most annuals will bloom so much better than perennials, with the exception of scabious and certain members of the dianthus family, but as with all soil types, certain plants prove more satisfactory in various soils; annual carnations, asters and scabious being particularly suited to a chalky soil. The great value of the annual is in its ability to give a good account of itself under the most varied and difficult soil conditions, and for the new yet unprepared garden no plant will prove more colourful nor more inexpensive.

Chapter 2

ANNUALS; THEIR TREATMENT

The hardiness of annuals — Methods of raising the plants —Climatic conditions — Soil and Situation — Preparation of Soil — Planting for maximum effect.

ANNUALS, as we have seen, may be divided into those having a hardy constitution and those which may be classed as being only partially hardy. There are certain plants more hardy than others even when they still may be called hardy, whilst amongst the half-hardy varieties there are some which may be called much more tender than others in this section. The use of all those plants will depend upon climatic conditions, soil and situation of the garden and upon the methods available for raising the plants. Biennials, which are sown early in the summer of the year previous to that in which they are to bloom, may all be classed as being quite hardy, even in the most exposed gardens, with the possible exception of the annual scabious, which though strictly a biennial, should in the more exposed gardens be given half-hardy treatment, for it may not be sufficiently hardy to withstand a severe winter unprotected.

Few gardeners realise that many plants which are given biennial and even half-hardy annual treatment are really true perennials. Amongst these is the forget-me-not (myosotis), used for spring bedding and which generally is treated in the same way as the spring flowering biennials; and the antirrhinum which is always treated as a half-hardy or hardy annual, for in this way it produces the best results. This surely is the only method of determining into what category the various plants should be placed, for all plants must be given the treatment to which they best respond.

Methods of raising annuals

In the case of antirrhinums, though really perennial, they will be found to give a more vigorous display if treated as hardy or half-hardy annuals, but exactly which treatment they are given must be decided by the means available for raising the plants. Those who have a heated greenhouse will find that a colourful display may be obtained from plants raised in gentle heat early in the New Year. Where a cold frame is available then the plants may be treated as biennials, being sown in boxes in the open during July, the plants being pricked off into the cold frames during early September in which they are wintered.

Though not nearly so popular a method of rearing plants as sowing in a warm greenhouse, the use of a cold frame and providing the plants with somewhere between hardy and half-hardy treatment has much to recommend it. It is a method by which not quite hardy plants and those not really tender may safely be brought through the severest of winters in the most economical way. A cold frame may be inexpensively made, yet it is general for those who have no greenhouse to sow their annuals directly into the open ground, the hardy annuals in early spring, the half-hardy plants in early May with the result that though they are sufficiently accommodating to provide a colourful display, they will especially in northern gardens, be very late to come into bloom and there will be no continuity from the spring bedding display.

With those plants that are sufficiently hardy to withstand winter conditions under cold frames, those reared in this way will prove much superior to those raised in heat, for the plants will benefit from being able to store up a concentration of energy through the long winter months in much the same way as do bulbs when in the plunge bed. Then as soon as planted out, the plants will grow away rapidly and come into bloom several weeks before those raised in heat, or those sown in spring where they are to bloom. And having harnessed the maximum of energy the plants are able to give a vigorous display over a greatly extended period.

But this method can only be carried out if one decides upon those plants required for bedding or for cutting whilst the beds are in full bloom. The average gardener prefers to await the New Year before the seed order is sent out and thousands defer buying their seeds until the spring, with the result that the annuals never attain the glories of which they are capable, not having time to build themselves up before they come into full bloom. biennials demand a long season in which to mature and are given it, though in many cases the plants should be sown earlier than is general. All annuals, however, are expected to come into bloom in the shortest possible time. Some will do so and these are those valuable gap-filling plants, the night-scented stocks and the coloured flax (linum), which make only small plants and which come quickly into bloom, sown in much the same way as mustard and cress. They are hardy annuals but very different in their habit to an equally hardy annual such as the marigold or calendula which makes a robust, almost bush-like plant and bears a continuous supply of large, many petalled blooms over a long period. To expect the plants to give of their best if given the same treatment as the night-scented stock with its tiny blooms, a plant completely lacking substance, is to bring about considerable disappointment.

Perennials may, with one or two exceptions, namely the scabious and pyrethrum, all be planted in much the same way in autumn, from late October until the year end when severe frost makes the ground too hard to work. With annuals, however, almost every one should be given the cultural treatment to which it best responds; in other words, each plant should be treated on its merits.

Climatic conditions

But in one's selection of annuals to grow each year much will be determined by soil and situation, as well as by methods available for raising the plants.

To those fortunate gardeners situated in the mild, moist climate of the south west, those gardening west of a line drawn from Bristol to Bournemouth, will be presented with few problems in raising their annuals. All those of a hardy nature may be sown in the open in early September and will winter outside quite unprotected, whilst those less hardy may be wintered in a cold frame, cold greenhouse or under cloches, few varieties being raised in the heat.

Again, with the early spring sun able to warm the soil several weeks before that of other parts of the country, both hardy and half-hardy plants may be sown in the open ground early in March when they will come into bloom even before those living in the north will have sown their less tender varieties. Annuals here present no difficulties, they will give of their best from an open ground sowing.

Those who garden in the less favourable parts of Britain must give more careful consideration to the requirements of the individual plants. The grower of cut flowers situated in East Anglia or along the East Coast, where the winter climate is less troubled by frost than those inland situations in the north and Midlands, will find it possible to make an early autumn sowing of the hardiest plants of the cut flower annuals with the expectation of the plants coming through the winter with few losses. Here larkspur and nigella may be sown in the open ground, whereas in areas troubled by dampness and frost, they would prove a better proposition if sown in early spring. But in this way the plants would come much later into bloom and may not prove profitable from a spring sowing, so it may be better to concentrate on other plants as an alternative source of income. Or cloches could be used to give the necessary protection.

Soil and situation

Site situation also calls for some consideration. The taller growing larkspur and scabious would either require staking, or suffer severe damage where growing in a windswept garden. The alternative is either to grow plants with a more dwarf habit or to provide some protection, this by way of a hedge or by wattle hurdles, which will lend privacy to one's garden in addition to protection. Few plants will grow well in a windswept garden and some protection is the first essential to a successful display of annuals.

Gardens with a northerly aspect as against those enjoying a more favourable southerly aspect, cannot be changed by artificial means so we have to select those plants which will succeed in colder and more sunless positions. Zinnias and verbena would not be a success for they like a soft warm, sandy soil, but mignonette and convolvulus, myosotis and pansies would be perfectly happy. The same may be said about the various soils. All annuals and biennials like a well drained soil, preferably one containing some humus though a dry, sandy soil always grows better annuals than a moist, clay soil. In the latter, calendulas and campanulas, clarkia and antirrhinums will all give a good account of themselves. Whilst the verbena, the eschscholtzia and the mesembryanthemum definitely prefer a sandy soil.

Then there is the question of a chalk soil which is generally a dry, shallow soil. Here the annual lupin and the ursinia are never happy, but the tagetes, French marigold, the petunia, and all classes of stocks and asters bloom remarkably well, and of course all members of the dianthus family.

Though aspect cannot be changed, the soil may be brought into that state in which almost all annual plants will flourish. It has been said that they are the most accommodating of all plants, producing a colourful display in almost all soils even if the plants are not always seen at their best. But those who would take some little care in the preparation of the ground will be astonished at the increased vigour and freedom of flowering of the plants.

Annuals planted in those circular and rectangular beds used in our corporation parks, and in the famous country house gardens still kept in a state of preservation, will be planted into beds which have been specially prepared for many years and will lack nothing in their cultural requirements. But most annuals grown today are sown or planted into small, modern gardens which are either almost unmade, where the soil possesses little fertility, or they are planted into those old town gardens where humus is completely lacking and where deposits of soot and sulphur have, through the years, made the soil quite sour.

Preparation of the soil

To take the new garden first. Here the soil is often of a heavy, clay nature, the top soil having been removed, either sold or used elsewhere, when the foundations of the house were being made. First it will be necessary to provide drainage, making the soil as porous as possible by incorporating crushed brick, often to be found about the garden of a newly built property, mortar, gravel or grit, and coarse sand. Each or any of these materials should be thoroughly worked into the soil to open up the clay.

Then dig in as much humus as it is possible to obtain. This by way of decayed leaves, peat, spent hops, wool shoddy readily obtainable by northern gardeners, and decayed manure , though it is realised that this is almost unprocurable by the townsman. A light sandy soil will also require humus, though not for the same reason. Here it will be required to hold summer moisture which will make artificial watering during a dry season unnecessary. Though annuals will prove free flowering in a dry, light soil, their freedom of flowering during dry weather will be greatly enhanced by the addition of humus for the plants to get their roots into. Plants kept too dry, will finish flowering before their time.

But though annuals enjoy some humus in the soil, too much nitrogenous manure must be avoided, otherwise a tall, lanky plant will result and the plants will make an excess of foliage to the exclusion of bloom. Should the soil be completely lacking in humus, then the plants will make little growth and though they will smother themselves in bloom, they will be short lived. To strike the happy medium is ideal for most annuals, though there are exceptions to this, certain plants being happier in a soil almost completely lacking in humus. This is where the annual proves its versatility, there being some plants for every soil and for all situations. But for annuals and biennials in general, provide them with some humus, but take care when using nitrogenous humus such as shoddy and farmyard manure.

Though necessary for perennials, the annual requires only very small quantities.

A soil of a sour nature, frequently found in town gardens should be given a liberal dressing with hydrated lime during the winter, for few annuals will bloom at their best in an acid soil. Town soils should be dressed with lime on alternate years. The addition of humus, unless the soil is of a clay or a sandy nature, will not be necessary here.

Lime should also be used in quantity to open up a sticky, clay soil, that is to be found around a newly built house or estate. Caustic lime, which has not been hydrated, is by far the most effective form, for the action of the moisture in the soil will generate considerable heat from the lime, causing it to expand and break up and in so doing will break down clay particles in the soil. Caustic lime should be applied to the soil at the same time as the drainage materials and this should be done in late autumn before the soil becomes too sticky. If left until March, when the humus is worked in, the soil will, after the action of the lime and frosts, be friable and well able to grow most annuals.

A chalk-laden soil, generally a shallow, hot, dry soil will respond to liberal applications of humus-forming materials, also by green manuring, digging in rape when about two inches high. Such a soil cannot be given too much humus, but here again the nitrogenous manures should be used sparingly. Peat is an excellent material to use in a chalky soil, for it tends to neutralise somewhat the alkalinity of the soil.

Remember to select a position of full sun for the annual display, they are never happy too near tall trees, nor in those gardens where adjoining property casts its shadows to shut out almost all the summer sunshine. But even where this occurs there are certain plants which will put up a brave show. With annuals there is something for every garden.

Planting for maximum effect

Annuals may be used in various ways, for edging, for carpeting, for spring and summer bedding, planted about the border or shrubbery, and in rows or beds for cutting, the range of plants being so extensive that a garden may be a thing of beauty by the use of the annuals alone. But in whatever manner they are used, only will they reveal their full glory when planted in mass. The odd plant dotted here and there about the garden will appear insignificant, lost amidst the foliage of the garden, but plant the same plants in beds or in clumps of a dozen or so and the effect will be one of great brilliance. This does not mean that the plants should be crowded together in such a way as to give them little room in which to develop in their natural form. This would be a waste of plants and would not enable them to give of their best; but to plant in groups correctly spaced, will be to ensure that the plants reveal their maximum brilliance. In this respect the annual has moved far from its reputation of a century ago, when Robert Thompson, then in charge of the Royal Horticultural Society's gardens at Chiswick, wrote, "a large proportion have a weedy appearance and are but ill-suited for the decoration of the flower garden". Perhaps that was so when Thompson wrote his *Gardener's Assistant,* but since that time the annual has been developed out of all recognition. Today it is a plant which for its versatility, freedom of flowering, ease in culture and its compact habit, enables it to hold its own in the company of any modern garden plant.

Chapter 3

METHODS OF SOWING THE SEED AND RAISING THE PLANTS

Dividing the plants into groups — Use of annuals to suit soil and situation — Autumn sowing — Value of a cold frame — Raising the plants — Sowing in heat — The use of a hot-bed — Value of cloches.

PERENNIAL plants are propagated either by division of the root-stock or from cuttings, for only in this way will a true stock be obtained. Annuals however are almost entirely propagated by sowing the seed, the modern strains being most reliable in producing a plant which will bear a bloom differing in no respect from the parent variety. Raising the plants from seed is both inexpensive and satisfying, it is in fact the easiest method of plant propagation, a child or one not possessed in any measure with so-called 'green fingers' being able to manage the sowing and raising of most annuals. Most gardeners will recollect their first tiny gardens where was sown Virginia stock and marigolds from vividly coloured packets and their joy at the successful germination and flowering. There is still the same enjoyment to be obtained from the sowing of annuals, but there is far more to it than sprinkling about the garden a packet of seed in spring.

For sowing purposes, annuals may be divided into five distinct groups:

(a) Those plants having so hardy a constitution as to permit them to be sown in autumn and wintered unprotected in the open. The same plants may also be sown in early spring as soon as the frost has left the ground but are better given biennial treatment. These plants may be called 'extremely hardy' rather than just 'hardy'.

(b) Then there are those which are quite hardy and may be wintered under a cold frame or under cloches in the north, though in the south-west they would come to no harm if unprotected. The same plants may be sown in gentle heat in a greenhouse early in the New Year and planted out late in spring after hardening.

(c) There are also various plants which may be classed as being only partially or half-hardy and which may either be raised in a heated greenhouse in the New Year, or sowing may be delayed until the coldest weather has passed, the seed being sown in a frame or greenhouse or even under cloches in early April, for in any case the plants cannot be set out until all fear of frost has departed.

(d) There is yet another group which by their nature and habit possess a shorter flowering season than other annuals and which respond best to sowing directly in the position in which they are to bloom. Here, in comparison to the calendula, in bloom from an autumn sowing from May until October, the viscaria and Virginia stock, flowering during July and August has but a shorter season and however earlier the seed may be sown the season cannot be prolonged. There is thus no reason for sowing in any other way than directly into the place where the plants are to bloom.

Into this group may also be placed all those plants which resent transplanting and which are better sown where they are to bloom. Like those previously mentioned in this section, many give a display of richness over a relatively short period, and even if they were tolerant of transplanting there would be little point in doing so. These plants are generally used for sowing about a border or shrubbery or in those odd corners about the garden which are devoid of colour. They may also be used on a rockery or for making an annual border. They are indispensable members of the race of annual plants and are possibly used in greater numbers than any other plants.

(e) Into this last group falls those plants which though not correctly annuals, are allowed to bloom for a single season and so are treated as such and must be considered an essential part of any book devoted to annual plants. These include the biennials (bi-annuals), and certain plants really of perennial habit. After all the antirrhinum is a true perennial, but it has become so much a part, possibly the chief part, of the summer display of annuals that it could not be separated from those annuals used for summer planting.

Let us take each section in turn and discuss its merits and requirements.

Annuals to suit soil and situation

In the previous chapter it was mentioned that most annuals could accommodate themselves in various soils and give a colourful display if those soils had been brought into a reasonable state of fertility, but it is not possible to change the situation of one's garden and one's climate. Thus if we live in the colder regions of the north, we must cultivate those plants well able to withstand these conditions. But our selection of plants must go further than this, for though almost all annuals, with the possible exception of zinnias, will tolerate northerly conditions, several annuals will only do so if raised in heat and are planted out late, and then will have but a short season for flowering. Here I am thinking of the verbena, which is at its best in the north only during late summer and early autumn. The salpiglossis is another and we may possibly include the schizanthus, though it blooms earlier and is rarely used for bedding. Each of these plants, as lovely as they are, will be better left to those gardens south of a line drawn from Chester to the Wash. Always use those annuals which may be relied on to yield a display of bloom over as long a season as possible in one's particular garden. Just because one may have noticed a particularly fine display in a garden in Devon does not mean that a similar show can be expected from the same plants growing in Derbyshire. Upon the realisation of this will depend the methods of raising the plants and the selection to be grown.

In the first group (a), sowing and care of the plants presents no difficulties. Into this group come many of the annuals used for cut flower purposes, larkspur, calendulas, annual gypsophila, nigella (love-in-a-mist), cornflowers and Shirley poppies, all of which may be sown in a prepared bed in the open ground.

It is important to select a sunny position, but one sheltered from the prevailing winds. Especially is this vital where those gardens may be troubled by cold winds which blow off the sea or across low lying land which will cause 'burning', or browning of the foliage. The seed may be sown where the plants are to bloom, such as larkspur and nigella which seem to resent transplanting, or as in the case of calendulas, the plants will most likely be moved to beds in spring where they are to bloom. Plants which are to be moved may be sown in dappled shade, such as provided by an orchard, but those sown where they are to bloom must receive ample sunshine.

The same remarks will apply equally well to those plants of group (e), which will most often be moved to beds for spring flowering. Though they will be sown at a different time, the soil will be given exactly the same preparation and methods of sowing will be similar, for the annuals of group (a) will be given what amounts to biennial treatment.

Autumn sowing

Annuals sown in autumn should have made sufficient growth to be able to withstand a hard winter, yet the plants themselves must be grown 'hard'. A soft plant making excessive foliage will be 'burnt' by cold winds, so guard against a soil containing too much manure. Soil containing a little old mushroom bed manure, some spent hops, or peat will prove ideal provided it is well drained and friable. Soil from which winter moisture cannot be drained away will prove quite unsuitable for the autumn sowing of annuals.

The seed should be sown early in September so that the plants will be just the correct size by November. In the north it is advisable to sow the seed towards the end of August for the plants will make little growth after the end of September. But first bring the soil to a friable tilth, containing few stones. Where the plants are to remain in the sowing quarters to flower, it is advisable to work in a small quantity, say two ounces per square yard of bone meal which will release its food value at a time when the plants are making growth late in the spring. To help the plants through winter, the soil should be given a one ounce per square yard dressing with sulphate of potash. This will build up a 'hard' plant and the potash will also increase the quality and richness of colouring of the blooms. It is important to sow in a soil which is not short of lime and which should have been applied the previous winter or well before the seed is to be sown. Also allow the soil to settle down and consolidate before sowing. In a loose seed bed the tiny seedlings may collapse owing to their roots not being in full contact with the soil. Some gardeners go so far as to roll the ground before sowing, but this should be done only where the soil is light and sandy. It

ween the rows and the soil slightly drawn up on either side of the plants as a protection against cold winds.

These extremely hardy annuals may also be sown in prepared seed beds in early spring, early in March in the south, at the end of the month in the north; but the plants will be a month later than autumn sown plants to come into bloom. A spring sowing is often made of those plants having a fairly short flowering season, so as to extend the flowering period right through summer. Also if gardening in a cold, exposed position and in a heavy clay soil, there may, in a severe winter, be considerable losses from an autumn sowing. As a precaution or as an alternative, a spring sowing should be made.

The second group of plants (b), those requiring glass protection through winter, antirrhinums, sweet peas, Brompton stocks, will all winter quite safely in the south west if unprotected, though the plants should be covered elsewhere. Here again it must be realised that these plants will produce an earlier and more vigorous display from autumn sown seed than if sown in heat in the New Year, which is an alternative method.

Value of a cold frame

There are several ways of providing protection, either by a cold greenhouse, frames or cloches, but of them all, frames are to be preferred for they afford more protection. The heated greenhouse will be in use for raising tomatoes and the more tender annuals.

For those who wish to raise their own annuals both for garden display and cutting, a frame will be indispensable not only for wintering certain annuals but for hardening off those plants raised in heat. The ideal is to have two frames available, their size depending upon the size of the garden and the number of plants needed to stock it. In one frame is placed those boxes of plants sown in autumn, and when removed early in April for planting out, seed may be sown of the half-hardy annuals; the other may be kept for hardening, or if half-hardy plants are grown in heat as they frequently are, then the one frame would be sufficient. the boxes of half-hardy plants are taken from the greenhouse during April to replace those plants which will have occupied the frame throughout winter. But the great value of a hotbed must not be forgotten as a much more inexpensive alternative to the heated greenhouse and this again is where the cold frame comes in useful.

A cold frame may be quite cheaply constructed by using old railway sleepers which have been creosoted and will be free from disease, or stout boards may be used. As an alternative bricks or breeze blocks are equally suitable. They are durable and should remain free from pest or disease.

It is advisable to use a new frame light, for those which are second-hand may so easily break with the weight of glass when moved. Also a light long exposed to deposits of soot will prove difficult to clean thoroughly and the glass may have to be replaced in any case. It is essential that where any form of glass is being used to accommodate plants, it must be kept thoroughly cleaned otherwise the plants will become drawn and weakly through lack of the rays of sun and light. This is of particular importance where plants are being grown under glass in an industrial town.

The size of light should be 5 ft. x 4 ft., certainly no larger otherwise it will prove too heavy for the lady members of the household to move and it is they who so often must attend to mid-day watering, shading and ventilation during spring.

To make a frame capable of accommodating plants in boxes, no particular care need be given to the soil unless the plants are to be transplanted to the frames rather than to boxes, or where seed is being sown directly into a frame rather than in boxes or in the open ground. In this case the frame soil must be well drained and should be friable, containing some moisture holding humus and yet at the same time should be of a porous nature. Some peat or leaf mould together with some coarse sand should be incorporated, then immediately before planting or sowin the compost should be gently pressed down. The plants are set out 3 ins. apart, planting them firmly.

Raising the plants

The seed of these less hardy plants will have to be sown late in August or early September either in boxes or pots of prepared compost, or directly into the compost of the frame either in drills 4 ins. apart or broadcast. Always sow thinly.

The compost for the seed sown in pots or boxes, whether under frames or in a heated greenhouse should be of the John Innes formula, as below, or one of the soil-less composts.

2 Parts loam, preferably sterilised
1 Part coarse sand
1 Part peat, of horticultural quality
1½ ozs. Superphosphate
¾ ozs. Ground limestone
per bushel

This compost or the soil-less will be suitable for the raising of all annuals from seed.

The seed of sweet peas and zinnias is often sown singly in 2½ in. pots, the plants giving a better display if they have no root disturbance. For boxes, kipper boxes or those specially made for seed sowing, being of a size easily handled and about 2 ins. deep, are suitable, whilst earthenware or plastic seed pans may also be used. All are equally suitable, but before sowing the seed make certain that the containers are quite clean. They should have been scrubbed before being filled with compost. The compost should have been kept quite dry and away from frost if being used during winter. It may be stored in an open shed, in a frame, or under the greenhouse bench. It may then be used as required.

The seed is sown thinly and just covered with compost which should be kept comfortably moist though not in any way saturated. The boxes or pans may be stood in a partially shaded position in the open, each box being covered with a sheet of clean glass, or they may be placed in the frames to germinate. When large enough to handle the seedlings are pricked out into boxes, each containing four dozen plants, or into frames where they remain throughout the winter, being watered only on rare occasions and given ventilation whenever the sun is warm. Excessive watering between

November and early March, whilst the plants are more or less dormant, will cause them to damp off. Ventilation is important for the plants must be grown as 'hard' as possible, requiring no coddling in any way for they are quite hardy even though better given winter protection in some parts.

Sowing in Heat

Those plants which may be classed as being only half-hardy, those of group (c), will be sown either over a mild hot-bed in early March or in a warm greenhouse early in the New Year. It is quite possible to sow seed in a cold frame in April or even in the open ground at the month end and whilst some degree of success may be obtained, the plants will come very much later into bloom. Though providing colour in the garden during late summer and autumn, they may be cut off by frost long before they have finished flowering.

Plants sown in heat or over a hot-bed

will be ready for hardening off by mid-April and may be planted out towards the end of May depending upon climatic conditions. They will come into bloom before the end of June, a month before those sown in cold frames and six weeks earlier than those sown in the open ground, though seed of the more tender annuals may be sown in the open in early spring in the south west and will still be in bloom before the end of June. But there are few lucky enough to garden in these favourable areas.

Seed sown in the warmth of a greenhouse will require more attention than those grown under cool conditions. They will require watering more often and greater attention must be given to ventilation. When watering seedlings, give the box a thorough soaking so that the moisture can penetrate to the roots. Frequent failure occurs through too light watering, the roots tending to grow to the surface in search of moisture.

High temperatures are not required, between 45 deg. F. (7 deg. C) and 50 deg.F (10 deg.C) being ideal; if too hot the plants will grow soft and lanky and will collapse with the first cool wind when planted in the open. Neither should the greenhouse be too humid. A close stuffy atmosphere will encourage disease and contribute to a weakly plant, so ventilate on all suitable occasions. To prevent risk of frost damage, all plants should be watered during winter before mid-day so that any surplus moisture will be dried off before nightfall.

Where growing in heat the plants should be kept as close to the light as possible. The greenhouse bench should be of a suitable height or failing this, shelves should be placed close to the glass and on these the boxes are placed.

It is important to prick off the seedlings as soon as they are large enough to handle otherwise they will become drawn. They are removed by means of a dibber, holding the seedlings carefully but firmly and making them quite firm in their new boxes. After transplanting the boxes are watered thoroughly.

The correct timing of the seed sowing is important, for the plants should not have outgrown the boxes before being planted out. If a plant has become too hard and short of food it may never completely recover and will remain stunted and rarely bloom to perfection. The plants should be of fresh appearance and short jointed when placed in the cold frame for hardening towards the end of April.

The use of a hot-bed

With the cost of heating materials now being so expensive, much greater use could be made of a hot-bed in raising half-hardy annuals. Indeed, those which may be considered the most exacting in their requirements may easily be raised in this way. The heat will be obtained free of all cost for after raising the the plants the compost may be used in the garden and will be no more expensive than if purchased for soil fertility purposes in the usual way.

A hot-bed is made by composting fresh straw stable manure or by artificial means, whereby straw thoroughly moistened, is decomposed by means of an activator. Dried poultry or pig manure may be added to enable the straw to attain considerable heat and so decompose rapidly.

Early March is the best time to make up a hot-bed, for by then the weather will be less severe and if correctly made the hot-bed will not quickly lose its heat. To make the bed, a frame should be made to the exact measurements of the garden light and soil to a depth of 18 ins. should be removed. The prepared compost is then placed in the frame to a depth of 15 ins. trodden thoroughly down to retain its heat and 4 ins. of soil is placed over the top.

where possible the soil should have been sterilised and be of a fine tilth. Into this the seed is sown, but not until the temperature in the hot-bed has fallen to 65 deg. F (19 deg. C). The seed will germinate quickly when ventilation should be given and by early April the seedlings will be ready for transplanting to another frame where they may be gradually hardened and planted out towards the end of May.

The value of cloches

Both cloches, glass or plastic and Ganwick frames may be used to cover rows of half-hardy annuals, indeed of all annuals. Their use will be to compromise between sowing in heat and in the open and is a suitable method for giving protection to many hardy annuals. But it must be said that several of the half-hardy annuals such as alyssum and lobelia require a longer period of growth before reaching a suitable size for planting out than cloche culture can provide. To be a success these plants must be grown over a hot-bed or in a warm greenhouse. But asters, cosmea and zinnias may all be successfully raised under cloches and of course those very hardy annuals sown either in autumn or the spring.

The culture of those plants requiring biennial-treatment is described in Chapter 5.

Those plants which are almost always sown where they are to bloom, should be sown in spring, depending upon the variety and as soon as the soil is free from frost and stickiness. Where the seed is very small, and this should also be done when sowing under glass, it is advisable to mix a little fine dry sand with the seed to enable more even distribution to take place. To prevent wastage of seed always sow thinly, especially the seed of those plants which cannot be easily transplanted. Where required for garden decoration, the seed is sown broadcast and gently raked into the soil which is kept moist during dry weather.

Chapter 4

ANNUALS; THEIR PURCHASE AND PLANTING

Not all have facilities for raising their own plants — What constitutes a good and poor quality plant — Hardening the plants — Planting out.

IT is realised that not all annual plants, biennials included, which are to be used for spring and summer display will be raised by those who are to plant them. They may not be able to find the time to prick out young seedlings and to rear them to planting size, with the need for specially prepared composts and with the necessary attention to watering and ventilation. Where the half-hardy annuals are being used for bedding, lobelia, alyssum, phlox Drummondii, etc., the gardener may not have facilities at his disposal for their correct cultivation. The use of only a few such plants, in the case of a small garden would not warrant the use of a heated greenhouse, or even of a hot-bed, or it may not be possible to use either.

For the purpose of supplying annuals for every purpose of garden display and for cutting there are specialist growers who advertise their plants through the gardening and daily press, and who in most cases can be relied upon to send through the post a well grown plant, well packed and which will give satisfaction if planted into a soil which has received average treatment. It must however, be understood as we have seen in previous chapters, that certain plants will succeed only under certain conditions. In a heavy soil and in a garden which may be troubled by cold winds, the zinnia would have little chance of success however well grown the plants may be.

What constitutes a good and bad plant

Not all nurserymen who advertise annuals in the spring and summer send out a plant worthy of their calling. So often they are 'drawn' and of a pale yellow colour, denoting that they have been raised in considerable heat, often sown late as a second batch, under almost forcing conditions. They are hurriedly hardened off and sent out far from being sufficiently hardened for the colder regions of Britain. As soon as planted out they will wilt under the first puff of wind and will take several weeks to revive and may never do so.

Or the plants may have been sown too soon and will possibly have remained in the boxes far too long awaiting transplanting, or even after having been transplanted. Some plants may have been moved to a none too suitable compost early in spring and they may have been hardened off in the normal way, only to be left in the boxes until well into June, when the plants will have become hardened and woody, the boxes being root bound, the individual plants lacking moisture and food. Here again the plants will have their work cut out to give satisfaction. They will be helped during a wet summer, or by a moist period immediately after planting when they may take in the much needed moisture, but if dry conditions should prevail after planting out they will never make suitable growth. They will come quickly into bloom and finish just as quickly.

The art of growing good annuals, whether used for bedding or for planting for cut bloom, is in correct timing, as in growing out-door tomatoes. The well grown annual plant should be of a rich medium-green colour, if too yellow in colour it will signify the use of too much heat and that the plant has not been correctly hardened. Such a plant should be rejected.

Besides the colour, the plant should be short jointed, sturdy and compact. Again, those long, lanky plants which flop over when planted should be avoided. If they cannot be replaced, they may be made to give a reasonable display by pinching out the main shoot to an inch or so of ground level after planting, to encourage the plants to form more compact, bushy growth. The modern strains of bedding annuals are now noted for their compact habit and there is little use the hybridist giving us plants with such valuable assets if the grower is to disregard their good qualities by poor cultural methods. The same remarks apply to those plants raised at home by various methods, the result should be the same in the end, i.e., a healthy vigorous plant by the time they are hardened out.

Hardening the plants

Hardening the more tender annuals is of the utmost importance, for no matter how carefully the plants have been raised all the good work will be nullified if the plants are not properly hardened before planting out. It should be said that those hardy annuals given the protection of a cold frame or of cloches for winter, will require almost no hardening apart from the removal of the glass by day from mid-March until early April with its replacement each evening. Those plants grown in heat will take three weeks for hardening, i.e., for the plants to become gradually accustomed to outdoor conditions.

The process of hardening must be gradual and it is no use being able to raise the tender annuals in heat if there are not proper facilities available for their hardening.

The plants will be ready for setting out at various periods depending upon site and situation and upon the varying degree of hardiness enjoyed by the plants. For instance, a plant such as the antirrhinum or marigold, raised in gentle heat will be able to be planted out three weeks or so before those of a less hardy nature so that hardening may be commenced earlier than for the more tender annuals. Likewise, those living in the south and south-midlands may commence hardening a fortnight earlier than northern gardeners; those in the south-west even a fortnight before those gardening in other parts of the south. Again, much depends upon the season. The spring of 1955 was known for its severely cold weather lasting almost until the end of May, so that hardening of tender annuals was delayed as long as possible and in the north hardening did not commence until the beginning of May.

There is always the desire to hurry the hardening operations in the natural desire to have the plants in bloom at the earliest possible moment, but it is disappointing to see plants which have been given so much care and attention during the first three or four months of the year, suffer a severe set-back simply because of one's impatience to complete the gradual process of hardening.

Hardening operations may be started even whilst the plants are growing in heat by the simple procedure of reducing the temperature, and providing more and more ventilation as the heat of the sun gathers strength. Open up the roof ventilators of the greenhouse but guard against draughts caused by a half-opened door and always open the ventilator windows so that they open against the prevailing winds. Cold winds of March can play havoc with a batch of healthy looking seedlings. And take care to see that the ventilators are closed and the temperature is increased by nightfall for a frost, even if only light, may also put an end to all the previous good work.

It will be about mid-April when the plants are taken from the greenhouse to commence the real process of hardening, a little later in cold districts, though much will depend upon the prevailing weather. Never fall into the trap of hastening the hardening process because a mild April is experienced. More often than not the following month may be severely troubled by frosts and cold winds. No matter how mild the late spring weather, it is not safe to give the plants their freedom until all fear of frost has departed which in the north will not be before June 1st, though it is safe to plant out all but the most tender of the plants during the latter days of May. In the north, however, the spring bedding display of biennials and bulbs will frequently, due to a late spring, occupy the ground until mid-June and the summer bedding plants must await planting. This is why it is important not to have the plants too long in the boxes for they will most likely have become root bound if they have to await planting until mid-June. This is where inter-planting proves its worth, for where polyanthus and violas and various bulbs are being used for the spring display, the summer flowering annuals may be set out between the spring flowering plants at the correct time and even slightly earlier, say by mid-May, for they will receive protection from the already established spring plants.

There is no better way of hardening the plants than by the use of a cold frame, however roughly it is put together. Should it be necessary to move the boxes to the frame whilst the weather is still very cold, additional warmth may be provided if soil, or better still, hot-bed manure, is heaped around the sides of the frame and the light is covered at night with sacking. If the heated greenhouse is required for an early crop of tomatoes it may be necessary to move the plants by April 1st. Normally the plants should go out towards the middle of the month, the lights being placed over the plants, ventilation at first being given during daytime. In case of frost it is advisable to cover with sacking in the evening for the first week or ten days.

The process of hardening off should be gradual, and towards the end of April ventilation may be given at night. Then during early May the lights may be removed half way and finally altogether, the material used to make the frame being sufficient to keep the frost and cold winds from the plants until ready to go out towards the month end. Generally it will be permissible to remove the lights during day time from the third week of April, but much depends on the prevailing weather conditions.

Where no garden frames are available, boxes of half-hardy annuals may still be satisfactorily hardened by using barn-type cloches or Ganwicks, the cloches being wide enough to take a standard size box, the Ganwicks wide enough to accommodate two boxes. The cloches may be covered with sacking or hessian canvas for the first nights and gradually more air may be admitted finally removing the top glass, using the cloches as if they were miniature frames. Or the cloches may be placed on their sides against the prevailing winds for the final hardening off.

Planting out

When the plants have been carefully hardened and kept growing on by careful attention to watering, keeping the soil sufficiently moist so as to prevent the formation of a hard, woody plant, they should be planted out at the first opportunity though if cold winds persist, it is better to wait a few days.

To complete the hardening, the

plants should be removed to the beds where they are to be planted, several days before they are to be set out. This will give them the opportunity to sample the conditions where they are to bloom, yet they will still receive the protection of each other. Water them well, thoroughly soaking the compost in the box so that it will adhere to the roots when the plants are removed from the boxes. This is why plants grown in one's own garden or purchasd from a local nurseryman in the boxes may be planted out with a large ball of soil on the roots. The plants will get away to a good start so much better than those plants purchased through the post, which will be almost devoid of soil upon arrival. However well grown the plants will have been, there is sure to be a check if the roots are so disturbed. Those plants whose roots are disturbed as little as possible will come into bloom considerably earlier than where root disturbance has occurred.

Planting should not be done until the soil is free from any stickiness . This will often prove troublesome when planting the hardy annuals in early spring, but rarely when planting out the tender plants towards the end of May or early in June. Then the weather is most likely to be dry and the soil lacking moisture so whenever possible await a shower of rain before planting out. If dry conditions persist then give the plants a thorough watering as soon as set out, otherwise they will tend to flag and may take several days to recover.

If the spring display is not ended by the time the summer bedding plants are ready to be planted out, the plants should be moved to a shady position where they are kept comfortably moist and away from the sun's rays. Moist peat should be pressed around the plants and if the boxes are also allowed to stand on a bed of moist peat this will keep the roots cool. Then plant them out at the first occasion. This is yet another point in favour of home grown plants which can be set out just at the right time.

Always plant firmly and whilst all annuals look their best when massed or planted in groups, too close planting should not be done, or the plants will tend to grow up rather than make that bushy-like growth which is the hall-mark of a well-grown plant.

Planting should be done with a trowel, removing the plants from the box and replanting into the beds with one action, so that the roots are not subjected to the drying rays of the sun a moment longer than necessary. On an average the plants should be spaced about 9 ins. apart, always planting between those first set out to allow them the maximum of room to develop.

Too formal planting of annuals should not be done, they are essentially informal plants and should, whenever possible, not be set out in straight rows by means of a garden line, as are geraniums for example, but at the same time the plants should be given the maximum of room in the beds.

Stopping the plants to encourage a more bushy habit may be done, though it must be said that the natural habit of the modern annual is one that is short jointed and bushy. However, many annuals will become even more neat and bushy if the leading shoot is pinched out when the plant is about 4 ins. high. Remove the growing point with an inch of stem; if less is removed it will perhaps grow again. The plant will commence to make fresh growth towards the base of the stem and so remain compact and bushy.

There has always been considerable discussion as to the exact time to 'stop' the plants, as this removal of the growing point is called. It depends much on the growth of the plants and no definite date can be given. But one thing is certain, it is not wise to 'stop' the plants at the same time as they are being planted out. Those plants obtained through the post and which may not have been 'stopped' should first be planted and allowed to become acclimatized to the soil for a full week before removing the growing points. Again this is where plants raised at home or obtained from a local nurseryman give the best results for they may be 'stopped' a week or two before planting. They will then have recovered from the shock before they are set out. A plant that is almost devoid of soil at the roots and which is pinched back heavily upon planting will suffer so great a shock as to take possibly the whole summer to recover.

After planting keep the plants growing on by watering whenever necessary. And keeping the ground free from weeds and the soil constantly stirred up to permit air and moisture to penetrate to the roots will make for a most rewarding display.

When the plants have become thoroughly established, and as they are about to come into bloom, occasional applications of dilute manure water will greatly enhance the colour and quality of the bloom and help to prolong the flowering period.

Chapter 5

BIENNIALS AND OTHER PLANTS REQUIRING SIMILAR TREATMENT

The requirements of biennial plants—Plants requiring biennial culture.

IN addition to the usual biennials which are generally grown for spring and summer bedding, there are other plants which respond to similar cultural treatment. Several may truly be classed as perennial and there are annuals, such as sweet peas and larkspur which also produce a better display if sown the previous year. But whereas these annuals are sown during the latter part of summer and would not be satisfactory if sown at any other time (see Chapter 3), these perennials which are allowed to bloom only the one season are sown during late summer, the true biennials being sown earlier. Bearing in mind that the half-hardy annuals are sown in heat during the early months of the year and that hardy annuals sown where they are to bloom may also be sown during spring, there is an all-the-year-round sowing season for plants which are to bloom during the same year.

It is most important to give those true biennials, e.g. wallflowers and Canterbury bells a long season; they do in fact require fully twelve months in which to grow before they come into bloom. Far too may biennials are sown late in summer, with the result that the plants will have made but little growth before winter. The vital period is summer and autumn for when November is at hand, growth will cease and the plants will make little more growth before coming into bloom in April. It is therefore essential that sowing is not delayed later than mid-May.

The wallflower, coming into bloom in April should in fact be sown towards the end of the previous April, and those plants which bloom later in the season are sown during May or June. The flowering period will be the guide to sowing, a full twelve months for growing being necessary for a plant to give of its best.

Requirements of Biennial plants

The perennial plant which is given biennial treatment, e.g. myosotis, sweet william, pansy, is generally sown early in summer, during June or at the beginning of July, the plants giving of their best from such a sowing.

All these plants, and the biennials,

will not only occupy the seed beds for as long as twenty weeks or more, but during that time will be making continuous growth, quite unlike hardy annuals sown at the end of summer and which have only a few weeks growing time before becoming dormant and remaining so throughout the winter. It is therefore of the greatest importance to allow the plants ample room in the seed rows and to transplant into prepared beds before the rows become overcrowded. To allow the plants to remain too long in the rows will result in a drawn, spindly plant, one which will form a large tap-root rather then the much desired fibrous roots. A plant with a large tap-root will not transplant anything like so well as one making a mass of fibrous roots, which will obtain the maximum of nourishment and moisture from the soil. It will also retain soil particles about the roots when the plant is being removed. A plant with a coarse tap-root will take several weeks to recover when moved and this really means that the plant will never recover completely to be the equal of that which suffers no set back upon being moved. A plant either moves continually forward or goes back and is never able to fully recapture the lost ground.

It must also be rememberd that seed sown during the early summer months will have to contend with long periods of drought; so that a soil well enriched with humus is essential. Also during dry periods the seedlings and young plants must be given artificial moisture.

But of the greatest importance is to thin out the seedlings as soon as large enough to handle and to transplant to prepared beds as soon as the plants are ready. The bed should be made very firm, again to encourage the formation of fibrous roots rather then forked or tap-roots.

When labour was inexpensive and readily procurable during the years preceding the First World War, biennials and more especially wallflowers, would be transplanted three or four times, once every month until finally bedded out with bulbs or planted into beds to themselves early in autumn. The result was a compact, bushy, fibrous rooted plant, well able to withstand a severe winter and enjoy a long flowering season. When transplanting wallflowers and *cheiranthus allionii,* it is advisable to pinch out the tap-root, though if the plants are moved when young and kept growing on through summer this may not be necessary. But with biennials and with wallflowers especially, far too many are used for bedding which are the result of sowing too late, too thickly and allowing the plants to remain too long in rows. The result is a woody plant, loose jointed and devoid of foliage, so essential in building up a vigorous plant.

Besides their use in beds to themselves, biennials are so valuable for carpeting or inter-planting beds of spring flowering bulbs. Yellow tulips of the Darwin or Cottage type provide a far more attractive display when carpeted with myosotis (forget-me-not) in either light or dark blue, or interplanted with scarlet wallflowers. Or to carpet a bed of single or double early flowering tulips with the now so popular winter-flowering pansies in contrasting colours, will provide a display of great charm. Both myosotis and pansies are really perennial but are generally treated as biennials and cleared from the beds when they have flowered in late spring. Not only are plants used in this way very attractive, but by their freedom of flowering, will enable a reduction to be made in the number of bulbs required. This will reduce the outlay of the spring display, an important consideration.

When purchasing plants from a nursery or through the post, make sure that they have been at least once transplanted, wall-flowers preferably twice, otherwise you will receive weedy looking plants direct from the seed drills and which cannot possibly give a good account of themselves.

Plants requiring Biennial culture

Biennials (b)
Perennials (p)

CANTERBURY BELL (b). Flowering at a height of nearly 3 ft. and in bloom through the early part of summer they are best planted on the herbaceous border, or against a wall or fence protected from the prevailing winds. To allow the plants a full twelve months to develop, the seed is sown early in June either broadcast or in shallow drills and should only just be covered with soil. Or sow in boxes covered with glass. The plants should be moved to prepared beds for growing on when large enough to handle. There are two types, the single and the double, or cup-and-saucer type, obtainable in blue, mauve, pink and white.

CHEIRANTHUS ALLIONII (p). This is the vivid orange Siberian wallflower, also obtainable in a rich yellow form, Golden Bedder. The plants grow in the rosette form, almost like sempervivums and do not require the same attention to detail as the ordinary wallflowers. It should be realised that the plants come into bloom when the wallflowers are finishing, and will continue right through summer. For this reason they should be planted either in beds to themselves or in groups to the front of the herbaceous or shrub borders. The plants may be allowed to remain in the seedling beds until early spring.

DIANTHUS HEDDEWIGII (b). Quite a recent addition to biennial dianthus grown in this country is the Japanese Pink, obtainable in the most vivid colours, brilliant crimson, salmon-pink, violet and almost black, both in the single and double forms. Flowering only 12 ins. high, they are ideal for bedding and should be given the same culture as for the other biennial members of the family. May also be grown as a half-hardy annual.

MYOSOTIS (p). This brilliant blue flower should be given a position of dappled shade, like that of the woodlands where it grows naturally. It is thus happy when used for planting at the base of apple trees or in beds that are partially shaded, and also when used to under-plant a bed of tulips. The seed is sown into a soil well enriched with some moisture-holding humus, such as peat or leaf mould and never at any time must the plants be allowed to suffer from lack of water. Sow in shallow drills in June and line the drills with peat before sowing.

For edging, Sutton's Miniature strain is most charming, the plants making little mounds only 3-4 ins high, obtainable in white, pink and blue. Growing to a height of only 8-9 ins in the bright Blue Perfection; the rosy pink, Carmine King; and the well-known, Royal Blue. The early flowering mid-blue, Express, grows to a height of 6 ins.

OENOTHERA (b). The Evening Primrose, so called on account of the large pale primrose coloured blooms and the delicate perfume given off on summer evenings. The plants grow to a height of 4 ft. and there are dwarf strains and remain colourful through the latter part of summer, the blooms soon dying but being replaced by new bloom almost daily. They should be planted towards the back of a border and about a shrubbery. The seed is sown in early July and if sown thinly transplanting is not necessary, the plants being set out into their permanent quarters in October.

PANSY (p). There are two distinct types used for bedding and which are given the usual biennial culture, the winter-flowering and the ordinary fancy or large-flowered pansies. Both are best sown in open ground drills during the latter part of June, the winter-flowering to come into bloom late in autumn. They will remain colourful until the following autumn when they are generally removed to make way for fresh plants. The large-flowered or fancy varieties come into bloom late in spring and remain colourful until the end of summer when replaced by bulbs or spring flowering biennials. It is advisable to treat all these pansies as biennials, the summer flowering may even be sown under frames in autumn or in March, and treated as annuals, though they will not come into bloom until late in summer. After a season's flowering the plants will have become straggling and untidy and will have lost their most vigorous period of blooming and so are best removed.

It is just as important with pansies as

with other plants given biennial treatment to thin out and transplant at an early age, otherwise the plants will become drawn and spindly. Transplant into beds of prepared soil, placing the seedlings 3 ins. apart and keeping them comfortably moist.

The winter-flowering pansies may be obtained in a variety of colours including Yellow Queen; Helios, yellow with a dark centre; Blue Boy, pale blue; March Beauty, velvet-purple; and the white, Snowstorm.

Of the summer flowering varieties, the rich apricot, Orange King; Ullswater, brightest blue; and Coronation Gold, rich yellow are attractive amongst the self colours. And of the Fancy pansies, Engelmann's Giant strain and Bath's Empress, the blooms obtainable in a wide variety of glorious markings are outstanding. The Felix pansies, with whisker markings radiating from the centre are an attractive novelty.

STOCK (b). It is the Brompton stock which is hardy and biennial, the seed being sown early in July to bloom the following early summer. The usual method is to sow in boxes in cold frames into which the seedlings are transplanted during August. The plants are wintered in the frames, watered sparingly and kept well ventilated for they are quite hardy but may suffer from cold winds in an exposed garden. The plants are planted out in beds late in March. Planting should not be delayed for they will soon come into bloom.

In the more sheltered garden the plants may be set out in beds in early October, allowing 9 ins. between the plants. Amongst the most colourful varieties are Empress Elizabeth, bright carmine pink; Cottager's Scarlet; pale mauve; blue; and white.

Rather later to come into bloom are the East Lothian Stocks, more often grown for cut flowers under glass. They require the same culture and may be obtained in the same colour range. Or they may be treated as half-hardy annuals, the seed being sown under glass early in spring or in heat in February.

SWEET SCABIOUS (b) This plant differs from those of the biennial section in that though a true biennial it is generally treated as a hardy annual and sown outdoors in the sheltered position late in August, or in a frame in an exposed garden. It may also be sown in gentle heat in the New Year, but it does require a long season to mature and with early spring sowing generally given to so many hardy annuals it is never a success, the reason being that it is really a biennial. It responds best if sown between the time usual for the ordinary biennials, early summer, and that when the hardy annuals are sown.

Given the half-hardy treatment the plants will come into bloom in early

August, a month earlier if treated as biennial. The deep ball-shaped flowers of the modern varieties make them very suitable for cutting for they have long wiry stems and are very long lasting in water. The flowers are produced on 2 ft. stems. Amonst the best varieties are Blue and Rose Cocade and the unusual Coral Moon.

SWEET WILLIAM (p). Treated by some as a biennial, by others as a perennial. It will however bloom early in summer the season after is is sown in May, so is included here. The seed may be sown in shallow drills in a partially shaded position or in a frame during June after the summer bedding plants have been hardened and planted out. Sow broadcast or in drills keeping the seed at all times moist. Sow very thinly for this is a plant which will soon become straggly and weak if not given plenty of room to develop. Transplant to a sunny bed when the plants are large enough to handle, preferably where they are to bloom.

Besides the well-known Pink and Scarlet Beauty strains, the Giant Auricula-eyed and Sutton's Harlequin strains are very attractive, ideal for bedding or for mixed borders. There are also dwarf strains growing to a height of only 8 ins., obtainable in pink and crimson and which are ideal for bedding.

SWEET WIVELSFIELD (b). Like the Sweet William from which it is a cross, this member of the dianthus family is also given biennial treatment. Growing to a height of 12 ins. and in bloom from June until the autumn, its Sweet William-like blooms are of almost every imaginable colour, and it is most suitable for bedding. There are single and double forms.

Seed sown either in the open or in frames in May, the plants being set out where they are to bloom at the end of summer. Like all members of the dianthus family this plant loves a well limed soil.

WALLFLOWER (b). Of all the biennials the wallflower seems the most difficult to grow well. The trouble is that it is always sown too late in summer, and is not transplanted as soon as it should be. The old country house gardeners would always transplant into a bed of fine soil 4 ins. deep over a layer of stone or slate to prevent the formation of tap-rots, but early and regular Fresh seed should also be used for large enough to handle.

A most attractive display may be obtained by filling a circular with the deep velvety red, Belvoir Castle in the centre, surrounded by Ivory Queen. Then a circle of Fire King, Orange Bedder, Eastern Queen and Cloth of God, the whole making a display of great splendour. Or they may be used for inter-planting a bed of late flowering tulips. About the garden they are essentially at their best when massed. The seed is sown in drills in the open as early in spring as possible, April not being too soon. For cutting, with their rich colourings and sweet perfume they are also indispensable.

Chapter 6

ANNUALS FOR BEDDING

The value of annuals for bedding — Making the beds — Annuals suitable for bedding.

WHERE the maximum of colour is required over the longest period, no plants are more valuable than those numerous annuals which possess a neat compact habit and which come into bloom early in summer. They may be used to inter-plant a bed of spring flowering perennials or bulbs, being planted out at any time from the beginning of May depending upon the soil and the situation and hardiness of the plant. In the small garden where concentration of colour is essential, there are no more suitable plants and none more easily managed. They may be grown from seed sown in the open late in summer where hardy annuals are to be used. No plant makes a more colourful display than the calendula, the Pot Marigold of olden times, used for bedding from a late summer sowing; or certain plants such as the antirrhinum may be wintered in a cold frame and the same frame used for raising more plants in early spring. When once the frame has been purchased the only outlay each year is a packet or two of seed and some care in the raising of the plants. There is no gardening more rewarding than raising from seed a colourful display of annuals for bedding.

Annuals for bedding have for long been popular. From early Victorian times the antirrhinum held pride of place with the geranium and calceolaria for the main summer display, whilst for edging, alyssum and lobelia have since those days never lost their popularity. But until more recent years there were few other suitable plants, generally because of the absence of a compact habit, whilst the colour range was not nearly so wide as it is today. As an example the now so valuable dwarf French marigolds, could until more recent years be obtained in either yellow or orange, flowering on 10 in. stems, and whether they turned out single or double was in the lap of the gods. Today, the dwarf French marigold may be obtained in more than a dozen varieties of both single and double forms, of various heights and with the blooms possessing all manner of unusual markings. With their long flowering season these dwarf marigolds have become one of the very best for summer bedding.

The same may be said of the antirrhinum, which today may be obtained for cutting, for border colour, for large and small beds and for the rockery, obtainable in a wide range of glorious shades and varying in height from 2-3 ins. to more than 3 ft. and where Rust

Disease has been experienced there are now a number of rust-resistant varieties.

Again, annuals are most valuable in that the modern bedding strains require almost no attention when once the plants have been set out apart from the removal of dead blooms. No staking is required, and when planted into clean ground and the complete clearing of the beds after flowering in early autumn makes this possible, there should be no need for any summer weeding for the plants will choke out all annual weeds. All of which is most important to the gardener with only limited time at his (or her) disposal.

Those fortunate gardeners possessing a heated greenhouse or who are able to make use of a mild hot-bed and a frame, may be able to raise a wider variety of plants and may obtain considerable interest from trying out new plants of a half-hardy nature which are not so commonly used for bedding. But with whatever plants are to be grown it will all depend upon soil and situation, those of a more tender nature being left to those gardens more favourably placed. For the less fortunate gardener there is still a wide variety of annuals available for bedding and many of the less popular plants may well reward the enterprising gardener.

The range of annuals suitable for use as bedding plants is now so varied that there is a great need for a little more originality. For instance, a welcome change for edging to the all too common white alyssum and lobelia would be the dainty powdery blue ageratum, the new lilac alyssum and the Tom Thumb antirrhinum. These plants are not yet easily obtainable from nurserymen but may so easily be grown by those with a cold frame or warm greenhouse. More co-operation may also be made with your local nurseryman by telling him of your summer requirements well before the New Year when he makes the first sowings.

He will be only too pleased to raise the plants of your choice.

Making the beds

In making a flower bed suitable for annuals, provided the soil is in good heart, little in the way of manures will be required. A good friable loam will ensure a display of brilliance from most annuals, but a 2 oz. per square yard dressing of bone meal forked in just before planting will help to maintain a vigorous display of bloom right into autumn, and if the soil is of a sandy nature work in some peat or leaf mould to retain summer moisture.

When making a new bed on a lawn or alongside a path, some care should be taken in making the bed so that the bloom is shown to the best advantage. When making a circular bed the size is first determined and this is marked out with a stake fixed in the centre, strong twine being used to make the diameter in four directions. It is then an easy matter to complete the circumference of the circle. The turf should be removed in sections and used elsewhere in the garden. The soil is then turned over, humus materials and bone meal being incorporated at the same time. The soil is dressed with lime and if possible is allowed to weather through winter. Soil is then raked towards the centre so that the centre of the bed is higher than at the circumference. This will enable plants to display their bloom to the best advantage.

The shape of the bed will depend upon the size and shape of the garden, a circular bed being most suitable in a circular lawn, the long rectangular bed being more suitable for a long rectangular garden.

Planting of hardy annuals is done during April, the more robust of the half-hardy annuals about mid-May, with the more tender plants being planted at the end of May or early in June depending upon the weather, soil conditions and if the spring display has finished.

Plants suitable for bedding

Hardy (h)
Half-hardy (hh)

AGERATUM (hh). A lot of work has been put into the Ageratum to provide compact, free flowering plants of uniform height in F1 strains with such types as Blue Mink, Blue Blazer, Summer Snow, growing to between 6 and 8 in. Blue Bouquet growing to 18 in. is useful for cutting. Seed should be sown in heat, the plants being set out early in June, 6 ins. apart.

ALYSSUM (hh). This is the Sweet Alyssum, at its best when used for edging a bed of scarlet geraniums or petunias. The plant, with its spreading habit is even more useful for carpeting and could be used more often with these two plants. Among the best forms are Snow Carpet; and the rich violet, Royal Carpet. Both are sown in gentle heat. They should be allowed 9 ins. in the beds when planted out.

AMARANTHUS (hh). Love-lies-bleeding is its more popular name, and which bears dark amaranth-red tassel-like flowers on 2-3 ft. stems. It makes a striking plant used as a centre piece for a large circular bed and is in bloom from Mid-July until late in October. Seed is sown in gentle heat early in March or in a cold frame about the same time.

A. caudatus atropurpureus is the best form, whilst there is also a form bearing palest green tassels and which makes a striking contrast if planted together.

ANCHUSA (a). The annual form of the more commonly known perennial is in every way a stiking plant for bedding. Sutton's strain of the variety Bedding Blue, makes a compact plant bearing masses of sky blue flowers through July and August, like large forget-me-nots. Seed should be sown in a frame as the plants may not survive a severe winter in an exposed garden.

ANTIRRHINUM (hp). This most poplar of all bedding plants is truly perennial but is best treated as a biennial, the seed being sown in a cold frame late in August, the seedlings pricked into boxes and wintered under a frame. In this way the plants may be planted out early in May and will come into bloom towards the end of June. the growing point should be removed to encourage a bushy plant. Another method is to sow seed in heat in January, or over a hot-bed in March, but the plants will be a month later to bloom and will not be so robust.

For bedding, the Intermediate Varieties are most popular growing to a height of 15 ins. Outstanding is the silvery-pink Malmaison; Guardsman, orange-scarlet with yellow lips; Eclipse, deep crimson; and Golden Queen. There is also a strain of more dwarf habit for bedding, the plants growing to a height of 9 ins. Amongst the best are Pink Exquisite, Yellow Prince and Cherry, but the colour range is considerable.

For those gardeners troubled by Rust Disease with this plant there is a new range of rust-resistant varieties, which grow to a height of 12 ins. Pink Freedom, with its almond blossom flowers; Golden Fleece, yellow and apricot; and Victory, buff and orange are outstanding.

For edging or carpeting, the Tom Thumb Sweetheart and Floral Carpet hybrids are delightful, having a large colour range and an almost prostrate habit.

ASTER (hh). The value of this plant and the range for bedding and for cutting is enormous, there being something to suit every taste. The seed may be sown either in gentle heat or in a cold frame in March. Germination and plant growth is rapid and the plants will be ready to go out at the end of May after the usual transplanting and hardening. A sowing may also be made under cloches early in April or in the open ground at the month end. The aster is a late summer flowering plant and so has plenty of time to make growth through the early summer before coming into bloom towards the end of July. As the plants are of a bushy habit allow 9 ins. spacing and they like a soil containing plenty of humus.

One of the best asters for bedding is the Dwarf Queen strain, wilt-resistant and making a short, bushy plant, obtainable in rose, crimson, white, blue and mauve. This is the first aster to come into bloom.

With their tiny button-like flowers obtainable in the usual aster colour range, the Lilliput asters, growing to a height of 15 ins. and with a branching habit, are excellent for bedding and valuable as pot plants. Of similar form are the new Remo asters with their very rich colours, whilst the dwarf Waldersee strain flowering over a long period at a height of 9 ins. is most attractive. All these button asters look most delightful when edged with the silvery-grey leaf Maritime Cineraria.

Also suitable for bedding are the Ostrich Plume asters with their long, twisted petals providing a shaggy appearance. The blooms are large and produced with great freedom. The cut flower asters are described in Chapter 7.

Like stocks, asters are prone to Black Leg Disease, which causes the stems to turn black near the base and the plants to collapse and die. Watering with Cheshunt Compound from the time germination has taken place will prevent an outbreak of the trouble. But as with stocks, grow the plants as dry as possible and transplant before the seedlings become overcrowded.

BALSAM (hh). This is amongst the most tender of the half-hardy annuals and though classed in the same half-hardy group as the aster and the African Marigold, it is not nearly so hardy. It must be raised in heat and should not be hardened until mid-May, being planted out early in June and then it should be given a sheltered, sunny position. Like the zinnia it does better in the south.

For bedding, the rose-flowered and Dwarf Bush strains are the best, the plants having a compact habit, not making excessive leaf, whilst the colour of the bloom is amongst the most charming of all annuals, being of various shades of pink, salmon and cream, at their best during August and September. A bed is most effective when edged with ageratum with its delicate pastel-blue coloured blooms.

CONVOLVULUS (h). The new range of dwarf bedding plants are a welcome addition to the range of hardy bedding annuals, far removed in their habit from that obnoxious weed with the lovely white tubular flowers. The variety Royal Ensign, with its bright ultramarine blue trumpets and the almost prostrate, Lavender Rosette, a lovely carpeting plant with its Lavender flowers and grey foliage should be more widely grown.

The plants may be raised in a cold frame from an early March sowing or may be sown where they are to bloom, thinned out and transplanted.

CORNFLOWER (h). *Centaurea cyanus* is its correct name and though for long popular for cutting, the dwarf forms are ideal for bedding, making bushy plants and growing to a height of only 10 ins. Seed may be sown in gentle heat in early February or in a cold frame in late summer to come early into bloom the following year. Or sow in a frame or under cloches in early March. Plant 9-10 ins. apart, setting out the plants in April.

The first of the dwarfs was Jubilee Gem, with its rich cornflower blue flowers.

KOCHIA (hh). The 'burning bush' of childhood days which gives off a vapour which will catch fire if lighted on a calm day. The shrub-like plant is attra ctive during summer with its feathery pale-green foliage, lovely when planted with geraniums and salvias, whilst during late summer and autumn its foliage turns a rich mahogany colour, equally as attractive. Seed may be sown in early May where the plants are to remain, but it is best raised in gentle heat from an early March sowing. Growing to a height of 2 ft. the plants should be given the same distance in the beds to mature. *K. Childsii* is the best species to grow.

LOBELIA (hhp). Though, like the antirrhinum strictly a perennial, the lobelia is always given annual treatment, the seed being sown in February, covering the seed only lightly with sand and covering the box or pan with a sheet of clean glass to hasten germination. Whilst being hardened, it is advisable to pinch back those shoots which tend to become straggly to encourage a bushy plant. The plants should be set out at the end of May or early in June when the spring bedding display has finished, the plants being used entirely for edging and should be spaced 5-6 ins. apart.

Though the navy-blue, Mrs Clibran, with its striking white eye is the more widely planted of all lobelias, the paler Blue Stone, is far more attractive planted with pink flowering plants, whilst the dark sea-blue Crystal Palace; the white, Snowball; and the uncommon crimson-red, Prima Donna, are all worth using by way of originality. Or three or four varieties may be planted alternately to produce a colourful effect.

MARIGOLD (hh). This is not the Calendula, but the tagetes species, the French and African marigolds which have now become so popular for bedding and have been improved more than any annual plant during recent years. Both may be raised in gentle heat from an early March sowing or seed may be sown mid-March either in a cold greenhouse or frame. The plants may be classed as being almost hardy and require no detailed care in any way. Both the African and

French varieties are ideal plants for a Northern garden, being planted out towards the end of May and remaining in bloom until November. The French varieties are ideal for small beds and especially for planting between a bed of polyanthus. With their vivid orange and yellow colouring no plants will make a more vivid or a more prolonged display.

The African marigolds are tall and dwarf growing, reaching a height of 2 ft. and more and down to 10 in. and so are useful for cutting or for planting in the herbaceous border or in large beds. Being of sturdy habit, staking is not necessary. The blooms are globular, almost as large as a cricket ball and freely produced.

Two of the most compact varieties are Orange Queen which does not exceed a height of 20 ins. and the golden-yellow, Crown of Gold, with its reflexed under petals. There are also may dwarf strains from 10 in. to 12 in. in rich colourings.

The French marigolds may now be obtained with the most arresting markings of mahogany and gold, the markings being quite unique amongst flowers. In the double-flowered section, Gold Laced, the golden flowers having an attractive edging of carmine-red; Harmony, rich orange, with the outer petals deep mahogany; and Rusty Red, the crimson petals being edged with orange, are outstanding and rarely exceed a height of 12 ins. In the single flowered Naughty Marietta, a tiny compact plant, the golden blooms blotched and striped maroon being most unusual. A whole host of new strains are now available.

MESEMBRYANTHEMUM (hh). These dwarf growing succulent plants, native of South Africa, should be given a sunny position and a dry, sandy soil. They will survive almost desert-like conditions and are ideal for bedding in the sandy soils of sea-coast gardens remaining in bloom throughout summer and autumn, the daisy-shaped flowers opening only when the sun shines on them. *M. criniflorum,* is also known as the Livingstone Daisy, and no annual has a more brilliant colour range with its blooms of crimson, pink, apricot and yellow and numerous art shades.

For the plants to come early into bloom the seed is sown under glass, either in heat, or under a frame or cloches in early March, the plants being set out towards the end of May. Or seed may be sown early in May where the plants are to bloom, though by this method flowering will be delayed several weeks.

NEMESIA (hh). Like the marigolds, the modern nemesia has been improved out of all recognition during recent years, the habit being more compact, more free flowering and remaining longer in bloom. Though classed as being half-hardy they are almost hardy, and like the marigolds are ideal for a northern garden. In sheltered gardens in the south the plants may be set out, after hardening at the end of April, with a second planting early in June to carry the flowering period right through summer. For an early planting seed is sown in gentle heat in early February, a month later for an early June planting; or seed may be sown in a cold frame early in March.

Both Blue Gem and Dwarf Orange may be used for edging, delightful as an edging to a path, for they grow only 8 ins. tall, the large-flowered or *Strumosa Suttonii* strain being used for the main bedding display. Their freedom of flowering and rich colours is difficult to equal and like the dwarf marigolds they are not troubled by wet weather. To make a bushy plant the growing point should be pinched out whilst the plants are hardening.

NICOTIANA (hh). This plan used to be confined to the mixed border, for growing to a height of 4 ft. and opening its blooms only at night it was planted only for its perfume. The new nicotianas not only make a dwarf, compact plant, they are extremely free flowering and what is more important the flowers remain open right through the daytime, a compensation for their having lost some of their perfume.

All the tobacco plants are given the usual half-hardy treatment and may be planted out from mid-May, spacing the plants 12 ins. apart and planting firmly. The varieties Crimson and White Bedder make compact, bushy growth and are seen at their best when planted together and edged with dwarf antirrhinums. Lime Green is a popular newer variety.

PERILLA (hh). This is a plant used for its rich purple foliage, making an attractive centre-piece for a large circular bed. *P. nankinsensis,* makes a compact plant 18 ins. tall. It is raised from seed sown in heat early February, for it is slow to germinate. Plant out early in June, allowing 15 ins. between the plants.

PETUNIA (hh). This is one of the more tender bedding plants requiring to be raised in heat and being planted out in early June. With their ruffled trumpets of the most intense colouring, great freedom of flowering and compact habit, no bed will make a more exotic display, nor one so colourful over so long a period. Though really perennial, the bedding petunia is always given half-hardy annual culture.

For outdoor planting the more compact dwarf bedding strains should be used, the more robust varieties being given greenhouse culture. Outstanding for bedding are all the modern strains in a bewildering variety of colour and form, many of them F1. hybrids.

The seed is sown in heat in early February, or for a later display towards the end of March in a cold frame, the plants being set out 9 ins. apart early in June.

PHLOX DRUMMONDII (hh). Quite the best strain for bedding is Sutton's Beauty, obtainable in every colour imaginable and having almost a prostrate habit. To obtain a bed of these plants at their best, the shoots should be pegged down over all the bare soil patches as they make growth. In this way the bed will soon become a veritable sea of bloom in the most glorious art colours.

Though the plants must be given a warm, sunny position they do like a soil containing some moisture-holding humus, particularly a little well decayed manure. The plants are given the usual half-hardy treatment and in favourable districts should be planted out in early May; early in June in the north. They come into bloom mid-July and continue until mid-October.

SALPIGLOSSIS (hh). No annual is more exotic with its large tubular flowers of the richest colours, with their velvety texture and veined in the manner of orchids. They grow to a height of 2 ft. and though when massed in a bed they hold each other up and require no staking, a bed should only be made up when sheltered from strong winds. The plants are raised in heat and planted out in early June or seed may be sown in a border early in May and the plants thinned to 8 ins. apart. This is a late summer flowering plant, the best strain being the Mixed Large-flowered.

STOCKS (hh). This is everybody's favourite on account of its freedom of flowering and its delicate perfume, there being no better sight than a bed massed with these plants in one of those delicate colours, chamois, rose, mauve or pale blue, for which the stock is famous. The plants are not too easily raised for they suffer from Black-Leg

Disease at almost any age and to ensure immunity it is necessary to use a sterilised soil. Water only when absolutely essential and then with Cheshunt Compound. The plants should be given fresh soil in alternate years when planted outdoors. Apart from this give the usual half-hardy treatment.

By far the best strain is Hansen's 100% Double, by which it is possible to remove any single flowering plants. This strain originated in Denmark. When the seedlings are ready for transplanting, having been raised in gentle heat, if the boxes are placed in a temperature of just under 45 deg. F. for 48 hours, it will be found that a small number of seedlings will have turned a dark green colour. These should be removed and destroyed for they would produce only single blooms. The others are returned to the greenhouse, transplanted and grown on in a temperature of about 50 deg. F. and the result will be 100 per cent doubles.

The plants should not be set out until towards the end of May, being transplanted 8 ins apart in the beds. At all times keep the young plants as dry as possible.

SWEET PEA (h). The introduction of the Little Sweetheart race (6-8 ins) of this ever popular plant as well as other dwarf strains such as Bijou (12 to 15 ins) means that it can now be bedded out. There are other semi-dwarf strains which do not need the 'scaffold' work as for the normal Sweet Pea, these being Jet-set and Knee-Hi. The plants may be sown in early autumn or in a cold frame in March in gentle heat and planted out in May.

TAGETES (h). This is one of those annuals which may be classed as hardy in certain districts, but in less favourable areas should be given half-hardy treatment. Of the same group as the French and African marigold, it is *T. signata pumila,* a tiny compact plant which is generally called the tagetes. Varieties are Golden Gem, Lemon Gem, Red Carpet and Sundance, all flowering at 6 ins and making the brightest of edgings.

VENIDIUM (hh). Natives of South Africa and liking a light, sandy soil and a position of full sun. At their best in a dry soil were their large daisy-like blooms with their striking black centres remain constantly in bloom through the latter part of summer and early autumn. For a small bed, the Dwarf Hybrids should be planted for they grow to a height of only 12 ins. they may be obtained in a wide variety of colours; ivory, straw, yellow and orange and intermediate shades.

Taller growing, reaching a height of 2 ft., is *V. fastuosum,* the large orange flowers held above the interesting silky-grey foliage.

The plants are raised in heat in March, hardened in the usual way and should be set out from the end of May, a week later in the north.

VERBENA (hhp). This delightful bedding plant is yet another of those tender perennials which is always given half-hardy treatment. In the north there is a tendency for the plants to come rather too late into bloom, being at their best during late August and through September. In the south the plants from a February sowing in heat, come into bloom at the end of July. Like the Phlox drummondii, the shoots may be pegged down to fill the bed with colour, their flat heads in red, pink, purple and white composed of numerous tiny flowers, each with a striking white eye, being most attractive, especially when used to underplant a bed of gladioli. The plant enjoys a position of full sun, but likes a soil containing some moisture.

ZINNIA (hh). This glorious annual, a native of Mexico, is one of those more tender plants. It likes a sunny position, a light sandy soil, and an absence from rain. In the north it is happy only in a summer where the weather, like that of 1955, is above average and then to see it at its best it should be given a seaside or country garden where the air is in no way polluted. Where it flourishes, the modern zinnia far surpasses all other summer flowering plants in the brilliance of its display with the rubbery texture and vivid colourings of the blooms.

Though growing to a height of 2 ft., the flower stems are so sturdy and erect that staking is not necessary. As the zinnia resents transplanting, seed is germinated in a heated greenhouse in March and as soon as as large enough to handle, the seedlings are transplanted into 2½ in. pots and kept growing. Later they are moved to 3 in. pots and hardened off during May. As for stocks and asters it is advisable to water only sparingly and then with Cheshunt Compound made by dissolving 1 oz. of Compound to 2 gallons of water. The Compound is made by crushing together 11 parts of Ammonium Carbonate and 2 parts of Copper Sulphate.

The plants should not be set out until early June, being planted from the pots containing a ball of moist soil. plant 12 ins. apart. They will come into bloom mid-July.

For bedding the Giant Double strain with its beautifully rounded blooms obtainable in rose, white, yellow, orange and scarlet and always looking their best when mixed.

Having larger blooms and being more globular in form are the Giant Dahlia-flowered zinnias, whilst the blooms of the Chrysanthemum-flowered strain have long shaggy petals. For a miniature bed or for edging, the tiny Pompon, Thumbelina and Peter Pan strains growing from 6 ins. to 1 ft. are most attractive.

Chapter 7

ANNUALS FOR CUTTING

Where to plant the cut flower annuals — Sowing the seed — Suitable cut flower annuals.

TO provide cut bloom for the house or for re-sale, annuals are valuable plants, extremely free flowering, hardy and most inexpensive to raise. Nor are they particular as to soil, being happy in any well drained loam and not requiring the copious amounts of manure as do chrysanthemums and many other perennial plants. There are annuals suitable only for cut flower purposes and which would, on account of their tall habit be out of place used for bedding, though many would be suitable for the annual border. But there are a number of dual-purpose plants, perfectly happy when used for bedding as well as to provide cut bloom. In this category come the zinnia and aster.

Cut flowers about the holding where they are grown for profit present no difficulties, they are an essential part of the enterprise and must be given a position which suits them, but to sow beds of annuals about the private garden just to provide cut bloom is not always easy. A pleasant idea is to use these plants to divide the kitchen garden from the flower garden, or even to divide the various vegetable crops. In this way they help to give each other protection from cold winds. For instance if a bed of larkspur or nigella is sown alongside a bed of mature Brussels Sprout plants which will remain in the ground until the early spring, they will provide excellent shelter. Cut flower annuals also fit readily into rotational cropping in the vegetable garden for they are at their best when sown in soil which has been manured for a previous crop. They could well follow a crop of lettuce after potatoes, the seed of the most hardy varieties being sown towards the end of August, or early September in the south. The plants will have made sturdy growth before the severe weather. To plant amongst vegetables means that the annuals will require no additional protection, otherwise twigs should be stuck into the soil about the plants. They should of course be sown where they receive some protection from the cold winds, especially those cold winds of early spring which cause the foliage to 'burn' just as new growth is commencing. These winds can cause more harm than frosts.

But as important as the provision of some protection is, all the cut flower annuals require a sunny position. To grow them in shade, possibly too close to a hedge, will cause the plants to grow spindly and weak and liable to suffer from damage during winter.

Where cloches can be used then this is the ideal method, for it is then possible to make autumn sowings of all the hardy annuals including those which are not quite so hardy as others. Not only are the plants given valuable winter protection, but they will come into bloom several weeks before the unprotected plants. This, to the market grower, will mean all the difference between large and small profits and to the housewife it will mean a valuable saving of her allowance at a time when there is little cut bloom about and flowers are expensive. If barn cloches or Ganwicks are used they may be kept over the plants for several weeks longer than where the tent cloche is used and which will have to be removed as soon as the plants make new growth in April.

Where one's soil is of a heavy, clay nature or one's garden is situated in a cold bleak district, it will be better to rely on growing the plants as annuals and not as biennials. This will mean that the plants will be later coming into bloom, but better this than to lose the plants. Those plants which may easily be transplanted may be sown in autumn and wintered under frames, or sown in frames in early March and moved to the open ground early in April. Where no glass is available this is also the best time to sow outdoors.

Too early sowing in the north and where the soil is heavy , will only mean that in a cold spring the seed will not germinate and may rot away, and unless sowing under cloches, it is advisable to delay sowing until the soil begins to warm in April. In friable soils in the south sowing may be done about March 1st.

Where cloches are being used to cover the autumn sown plants, they may be moved early in April to cover sowings of half-hardy plants so that a quick succession of bloom may be obtained for cutting in the quickest possible time. This would mean that the autumn sown annuals would come into bloom mid-May, the half-hardy plants coming into bloom early in July. If a second sowing of hardy annuals was made early in May, these would come into bloom when the autumn sown plants were finishing. Also if biennials were sown the previous May, these would provide bloom for cutting from late April, thus giving cut bloom for eight months of the year, for the May sown annuals would continue until mid-November. This method of sowing and planting for a succession of bloom may also be done where planting a border of annuals to provide both colour in the garden and bloom for the home. Such a border is excellent for growing cut bloom if it is judiciously cut so that the balance of the colour is maintained.

Sowing the seed

Sowing in shallow drills is more reliable than broadcasting chiefly because it is easier to keep the rows clean. The land for sowing, whether the seed is to be sown where the plants are to bloom or in specially prepared seed beds, must be quite free from weeds, especially chickweed, which is so difficult to eradicate without disturbing the plants. Farmyard manure, which is the source of so much weed infestation should not be used, but old mushroom compost which will be almost free from all weeds due to its attaining high temperature in the composting heap, is ideal. Where this cannot be obtained, peat also free of weed spores, is equally good, with spent hops an excellent substitute. The material which will supply the necessary humus must be well forked in in plenty of time for the soil to settle down before sowing. Humus is equally essential to retain moisture, whilst it is also important to prepare a fine tilth, air pockets caused by stones and lumps of soil being detrimental to the seedlings. Roots not in complete contact with the soil will rapidly dry out and die back. Thus a fine tilth and a compact soil is essential to the raising of a good plant.

Deep sowing is not advisable, drills 1 in. deep being correct and 8-9 ins should be allowed between the drills so that the soil may be kept clear of annual weeds by careful hoeing. Weeds too should be removed from the plant roots by careful manipulation and any disturbed seedlings should be pressed back into the soil, making them quite firm.

It is essential to sow the seed thinly for seed obtained from a reputable house will give an almost 100 per cent germination and overcrowding will produce a spindly plant which will never recover to produce the maximum of bloom. Even where the seed has been thinly sown it may be necessary to thin out the seedlings as soon as large enough to handle. Especially is this necessary where the plants are sown in their flowering quarters and any undue bunching together will again make for a disappointing display. Larkspur, nigella and cornflowers are the plants which resent disturbance and are always allowed to bloom where they are sown, in the more exposed districts being covered with cloches during the winter. Where winter weather is generally severe, sowing should be delayed until early spring, but then, even though cloches are used, flowering will be a month later than from autumn sown plants.

Those plants which will bloom better when transplanted, should be pricked out into prepared beds at an early stage so that they may become established before winter or they may be left in the drills until early spring. Sowing will determine this for seed sown late and taking rather longer to germinate than thought possible, would be better left undisturbed in the drill except for thinning. Much depends upon the situation and prevailing weather.

From the moment the seed is sown, the soil should never be allowed to suffer from drought. If the weather is dry, then artificial watering should be carried out whenever necessary. To allow the young seedlings to suffer from dry conditions at the roots will cause them to die back, which if only partially will mean that they will take several weeks to recover, with the ultimate delay in flowering time. For this reason, it is necessary to incorporate a considerable amount of humus into the soil, for soil which quickly dries out will not produce a really good plant.

The plants will require little attention after being thinned out in autumn, but as they make considerable growth in spring the taller growing plants will require staking. This is done by passing strong twine along the rows at both sides of the plants and which is held in position by stout stakes at intervals.

Before the plants make new growth in spring the hoe should be taken between the rows to break up the soil which may have become consolidated with the winter rains and snow.

After the plants have finished blooming, they should be cut away at soil level either with the shears or with a spade. The plants should then be chopped up and be allowed to remain on top of the soil for a day or two. They should then be dug well into the soil rather than burnt, or they may be moved to the compost heap. Most annuals are an excellent source of green manure, and especially where the soil is impoverished they should be dug in during autumn and allowed to decay over winter.

Suitable cut flower annuals

ACROLINIUM (h). This is a hardy annual which deserves to be better known. The seed is sown either in early September in favourable districts, or under frames or cloches in March. It is an everlasting flower growing to a height of 15 ins. and bearing delicate rose-pink flowers which are exceedingly long lasting when cut and dried early in autumn. There is also a double form, *flore plena,* and a white counterpart. When fully opened the bloom is cut with as long a stem as possible, almost at ground level. It is made into bunches which are hung suspended for a fortnight in an airy room, not a greenhouse, before being used for home decoration.

AGROSTEMMA (h). This a a hardy annual plant which blooms best from a sowing made in autumn or early spring where it is to bloom. The novelty Milas, bearing large flowers of a pleasing shade of soft lilac, on 3 ft. stems is the plant to grow for cutting. Sow the seed thinly and space to 6 ins apart. The bloom should not be picked until fully open when it will remain fresh in water for several days.

ASTER (hh). It is the China, or large-flowered aster which is so valuable for cutting. Seed is sown in gentle heat in March, or under cloches or in a cold frame in April to have the plants in bloom as quickly as possible. Where no glass is available, sow the seed in the open in early May.

Annual asters are great lime lovers, so work into the soil liberal quantities a month or so before the plants are set out at the end of May. The plants also like a firm soil, thus it should be allowed several weeks to settle down before planting.

The beds should be made 5 ft. wide to allow for ease in picking and as the plants are of branching habit, allow than 12 ins in the rows. Staking is not necessary.

CALENDULA (h). One of the easiest and most popular of all annuals. This is a plant which is generally sown in mid-September in cold frames or under cloches, the plants being pricked off in March into beds 5 ft wide, with the plants 8 ins apart. They will then come into bloom in early June, and like the cornflower will prove profitable well into August. Or seed may be sown in frames or under cloches in early March when the plants will commence to bloom in mid-July. Both methods are profitable though with all hardy annuals the early bloom catches the most profitable market. A sowing may also be made in May or early June and the plants transferred to frames, Ganwicks or cloches in August, a time when frames are not much in use, to come into bloom late in autumn when they will continue to bloom at least until the year end. Whilst it is not desirable to give the calendula too much nitrogenous manure or the plants will make too much foliage at the expense of the bloom, it does appreciate some humus, by way of peat or hops. This plant is never happy in a dry soil.

CARNATION (h). No annual is more pleasing for use as cut bloom about the home as the Chabaud Carnations of Raoul Martin of France. The blooms are fully double, have long stems and carry a delicious fragrance. They commence to bloom late in June and flower continuously until autumn.

This carnation should be treated as a biennial for best results, the seed being sown in a frame, preferably in boxes covered with glass, in July. The seedlings should be pricked out into the frames in September where they remain over winter and are planted out into beds in early April. Or the seed may be sown in gentle heat in boxes or seed pans in February and after the plants have been hardened they are planted out in early May. The best method is definitely to sow in mid-summer for then the plants come early into bloom the following season. For a sowing compost use one composed of loam, sand and a little peat and a small quantity of lime rubble. Where the plants are to be kept in frames through winter they should be dusted with sulphur once every month to prevent mildew.

The plants should be set 9 ins. apart in a position where they may receive full sunlight. They like a soil as light as possible containing plenty of lime and enriched with a small amount of cow manure.

CHRYSANTHEMUM (h). Bearing a great similarity to the calendula, the annual chrysanthemum, coming into bloom in mid-summer and flowering until early autumn, might be more popular than is the case, for they have a longer flowering season than most plants, last well in water and do not drop their petals. True, the plants do not winter well unless under glass, but this cannot be the reason for their lack of popularity with the public. Probably the new varieties of *C. tricolour,* will help to bring this flower into favour again.

The seed is sown under glass in September, preferably broadcast in a frame, and there the plants remain through winter for planting out 9-10 ins. apart in beds early in April. No staking is required . The bloom is cut when almost open.

CLARKIA (h). The clarkia is a most dainty flower when bunched and popular for the small vase in the small modern room. It does not like transplanting and for this reason should be sown in early April where it is to bloom. Nor does it winter too well except in the mildest districts, or in a walled garden where it may be sown in mid-September. If sown in spring, it should be cloche covered for it naturally blooms early. Plants from an autumn sowing will come into bloom towards the end of May, a most welcome time for there are then few flowers in the shops.

Seed is sown in drills 12 ins apart and the young plants should be thinned to 3 ins apart. Under good cultivation the plants will attain a height of almost 3 ft. and should be supported by twine on either side of the rows. dwarf stems are not wanted, but those of 18-20 ins in length and covered in bloom which is just opening. Liberal sized bunches should be made up.

CORNFLOWER (h). Always in demand, especially the pink shade which has now superseded the blue in popularity. The seed should be sown in early October in favourable districts and a month earlier elsewhere, possibly at the end of August in the most exposed gardens. Sow 2 ft apart and thin out the seedlings to 12 ins apart in March for the cornflower makes a tall, bushy plant. In early gardens the first bloom may be picked about June 1st from an autumn sowing. In less favourable areas, late June will see the first bloom. The best varieties for cut blooms are those of the Monarch Ball strains, long stemmed, extremely double, and very early flowering. Pink Ball, shell pink; Red Ball, rosy-red; and Blue Ball, devoid of the usual purple colouring are all valuable.

COSMEA (hh). Also known as cosmos, this is an elegant flower for cutting, at its best through late summer and autumn. It is late flowering for it does not readily transplant and must be sown early in May either under cloches or in the open. Sow thinly and thin out to 9 ins apart for the plants grow 3 ft or more and require some support. The best strain for cutting is the Early-flowering Crimson and the rose-pink Gloria.

GYPSOPHILA (h). This is another annual which likes a limestone soil or where this is not possible, quantities of lime rubble should be worked in before planting. A small sowing of this plant should be made in every garden, in any corner which receives the sun, for it is invaluable for mixing with other annuals especially carnations and sweet peas. Seed is best sown in early September, the seedlings thinned to 12 ins in early spring. The plants will produce a large quantity of bloom from mid-June until late September, which should be supported by twine fastened round the plants. The Monarch White strain is excellent and plant with it *G. elegans* Crimson and Rosea, both very lovely.

HELICHRYSUM (h). As they so readily transplant they are best sown either in gentle heat or in a cold frame in March, planting out early in May 12 ins apart. No annual possesses a wider colour range, the very double blooms being obtainable in all shades of rose and crimson, as well as gold, white and orange, all of which are enhanced by their attractive silvery sheen.

LARKSPUR. This is one of the most popular of all annuals today for it is a delightful plant for mixing with blue scabious and with other mid-summer flowers on account of its dainty, feathery habit. It is extremely hardy and not liking disturbance, the seed is sown early in September, in drills 2 ft apart where it is to bloom. Growing to a height of 4 ft. it should be given a sheltered position though it enjoys ample sunshine. The plants require staking against strong winds but otherwise no additional attention. The bloom must be cut as soon as it shows colour, whilst many buds are still green, for not only is the bloom enhanced by the green tint, but larkspur will quickly drop its petals when fully opened. The Giant Imperial and Double Stock-flowered strains are the most popular. of these, Los Angeles, with its salmon, rose tinted blooms; Rosamond, pure bright pink; Dazzler, rosy-scarlet; Blue Bell, bright mid—blue; and Sweet

Lavender, are outstanding.

NIGELLA (h). Love-in-a-mist is its more popular name on account of its misty green and blue appearance, produced by its green ruff. Because the plants from an early September sowing are in bloom by mid-June, it is always popular for apart from pyrethrums and Dutch iris there is little else available at that time. Growing to a height of 2 ft., the seed should be sown in drills 18 ins apart and very thinly so that in this case, thinning will not be necessary. Dappled shade suits this plant well and it likes a soil containing some humus. As the plants will finish flowering in about three weeks, it is best to sow again in late March to prolong the season and give mid-summer bloom, not quite so profitable, but still worth while. The bloom is cut just as it shows colour. It is better marketed locally rather than transported long distances.

RUDBECKIA (h). See Chapter 9.

SAPONARIA (h). It is the species *S. vaccaria* and the variety Pink Beauty which is so useful as a cut flower. It grows to a height of 2 ft. or more and bears its long sprays of deep pink flowers through late spring and early summer, when cut flowers are always in demand.

The seed is sown in drills 12-15 ins apart in August, thin sowing, making thinning unnecessary, nor should the plants be moved. In favourable districts they winter unprotected but it is advisable to cover plants in exposed areas with cloches or lights.

The bloom is cut just as it is opening, rather like gypsophila, taking a handful of stems and cutting off just above ground level.

SCABIOUS (H). The annual form of the scabious has for some reason never become as popular as the perennial, yet it possesses all the qualities of a top class cut flower, lasting well in water with the blooms of richer colouring than most flowers. The bloom is produced on 3 ft stems and is best sown where it is to bloom, the seed being sown late in August in drills 12 ins apart, the seedlings being thinned out to 6 ins apart in early spring. Any well drained soil is suitable provided it is not lacking in lime. The plants will appreciate some lime-rubble worked in before sowing.

Lovely for cutting is Blue Moon, of a glorious shade of sky-blue. The petals are large and give the appearance of Primula denticulata. Equally attractive is Cherry Red, well named; and Rosette, rose-pink, shaded with salmon, a bloom similar in shape to the Esther Read chrysanthemum.

STATICE (HH). Though more correctly of biennial habit, the well-known *S. Sinuata,* so popular for drying for winter decoration, should, like the helichrysum, be treated as a half-hardy annual. The seed is sown in gentle heat in early March, the young plants being moved to the open ground mid-May. Like all ever-lasting flowers they prefer a light, sandy soil. The plants grow to a height of 18 ins. and should be planted that distance apart.

The bloom is cut at ground level as soon as the bloom is nicely showing colour. The best varieties are Market Rose, Market Grower's Blue and Lavender Queen.

SWEET SULTAN (H). This is another of those hardy annuals which dislikes transplanting so should be sown where it is to bloom. Sow thinly in drills 9-10 ins apart late in August and thin out the seedlings to 5-6 ins. Should the weather be severe, wattle hurdles should be placed over the rows but otherwise the plants require no attention. They come into bloom in July and require no staking.

The bloom is cut when just opening.

ZINNIA. See Chapter 6.

Chapter 8

ANNUALS FOR THE BORDER

The Charm of a border of annuals — Annuals for dry soils — For cool damp soils — The treatment of border plants — Making the border — Care of the plants — Annuals in the shrubbery — Saving seed.

FOR long the herbaceous border has been firmly established in our gardens, yet the border of annuals, equally as colourful and remaining just as long in bloom, has remained almost entirely neglected. It has several advantages over the perennial border, one being that the ground may be cleared completely when the plants have been removed in late autumn so that all weeds may be eradicated. The ground will then require little attention during the busy summer months. Another advantage is that if labour is very limited, only those plants need be grown which do not require staking, whilst each year the border may be given a new dress, new varieties, possibly original, being introduced each year.

Annuals also prove highly suitable plants for the herbaceous border, especially for filling in all those spaces left by winter losses of the herbaceous plants, also where the herbaceous border is being built up slowly, selecting the best of the modern varieties when one's pocket permits their purchase. The business of making up an herbaceous border should not be done quickly without detailed thought being given to the colour scheme and the habit of the individual plants. It may take several years to complete and until then annuals should be used. For this purpose and for making a border entirely of annuals, or for using about a shrubbery which is not overcrowded, use should be made of those varieties which are generally sown where they are to bloom. These plants either resent moving, or they come into bloom so readily when sown in this way that raising under glass or in any way where transplanting is necessary, is never worth while. The plants are amongst the most valuable of all the annuals, indeed of all garden plants, for they may be sown to provide a brilliant display right through summer and autumn, and cost but a few shillings to beautify even the largest of gardens. during the present era of shortage of labour in the garden, their use is becoming continually more popular for they may be sown in a few minutes and require little more attention until the plants are removed in autumn. With most varieties neither staking nor tying is necessary. They will make a colourful display irrespective of the season, surviving drought-like conditions in any ordinary soil. Also their habit of producing new bloom whilst those which were previously formed are dying away, ensures a continuity of colour even without the removal of the dead blooms. I am thinking here of plants like the godetia, clarkia and the eschscholtzia.

Annuals for dry soils

Again, each of these annuals will give a good account of itself in a soil which may be classed as being anything but fertile, and this is where they are valuable in making a border in a soil which has received little attention, such as that found in the garden of newly built houses where the soil may take several years to become suitable for growing herbaceous plants and those plants demanding more exact conditions. In a soil entirely devoid of humus, which may be light and sandy or of a clay nature and which would prove suitable only for the rose to flourish, there are many annuals which will put on a brave display where they are to bloom.

For a dry, sandy soil these annuals will prove suitable and all may be sown where they are to bloom:

ALONSOA
BARTONIA
BRACHYCOME
CACALIA
CLARKIA
CORNFLOWER
ECHIUM
ESCHSCHOLTZIA
FRENCH AND AFRICAN MARIGOLD
LAVATERA
LUPIN (ANNUAL)
MESEMBRYANTHEMUM
NASTURTIUM
OBENOTHERA BISTORTA
PHACELIA
URSINIA
VENIDIUM
ZINNIA

Each of these plants will flourish under conditions which most perennials would make only stunted growth and remain in bloom for only a short period.

Annuals for cold, damp soils

Then there is the damp soil, where drainage is poor and where plants which would have to remain in the bed throughout the winter might die back through excess moisture remaining about their roots. Where annuals are sown this trouble is not experienced for the plants are removed in autumn after they have finished blooming. most of the annuals which are tolerant of a moist soil, and there is little fear of excess moisture collecting about the roots during the summer months, will also be happy in a northerly position, one where the sun's heat is at a minimum. Even in dappled shade they will put up a brave show. Here for such a soil are a dozen plants which may be sown where they are to bloom:

AGERATUM
CALENDULA
COLLINSIA
COSMEA
GODETIA
LIMNANTHES
LINARIA
LINUM (FLAX)
MIGNONETTE
NEMOPHILA
NICOTIANA
NIGELLA

In all fairness to these plants it must be realised that they will always bloom better in a friable loamy soil and in an open position, though they do not require any great amount of sunshine.

Treatment of border plants

An annual border should be made with just as much care and thought as to colour schemes, flowering time and the respective heights of the plants as given to the herbaceous border. There are numerous other plants to be used in the annual border apart from those plants which are generally sown where they are to bloom. Many of these which are generally raised under glass and are transplanted at some stage will also give a good account of themselves if raised from seed sown in the open ground. Some may be given hardy treatment sowing the seed in March or even in early autumn in the more favourable districts; others being given half-hardy culture, the seed being sown in late April or early May.

When sowing directly into the open ground sowing time will depend upon the condition of the soil and upon the situation of one's garden. There must be more than thirty annuals which may be sown in the open in early autumn, but any sowings should only be made if the soil is of a light, sandy nature which allows winter moisture to drain away quickly; and where one's garden is favourably situated. Those who garden south-west of a line drawn from

Bridgnorth to Brighton would in most years be able to sow the seed of these plants in autumn with the reasonable expectation of their surviving the winter. Elsewhere sowing of these hardiest of the hardy annuals should be delayed until early spring and until the frost has left the soil or until the soil had become friable and lost any stickiness.

Cloches have considerable value when sowing annuals directly into the open ground, for to obtain the best effect the seed should be sown in groups or clumps which would cover the area of a barn-type cloche. If a dozen such cloches were used about a small border, the seed of these hardy plants could be sown early in March even in the most exposed gardens, the result being that these plants would come into bloom before the end of May. Thus five weeks will be gained on those sown without protection and the necessary loss of time in waiting for the soil to become friable. The cloches should be placed in position early in March, a week or so before the seed is to be sown.

Towards the end of April, the cloches will be removed and used to cover the seed of half-hardy annuals sown in a similar way about the bed or border. They in turn will be removed at the end of May and may be used to cover plants of the more tender annuals which have been raised in heat and partially hardened off.

Plants raised under glass either from a late summer sowing and wintered in a cold frame, or raised in gentle heat may also be planted about the border when thorougly hardened. Thus with some plants coming more quickly into bloom than others, there will be, for the minimum effort, a continuous display from early summer until the end of autumn, twenty weeks of glorious colour. If cloches are not used planting will be delayed 2-3 weeks in all cases, the first plants coming into bloom early in June.

In those more favourably placed gardens, the use of cloches would not be necessary, for the seed of the hardy annuals could be sown in September, so that the plants will be large enough to survive the winter. The sowing of the more tender plants may be made early in April. It is in the more exposed parts that cloches prove so useful.

When making an annual border, it will be necessary first to dig over the soil, cleaning it of all perennial weeds and working in grit or coarse sand if the soil is of a clay nature. Then whatever humus is available should be worked in; seaweed, spent hops, leaf mould, a little decayed manure or some peat, taking care not to overdo the nitrogenous manures for the plants are required to produce colour not foliage. This work should be done in late autumn, the soil being limed and left to become weathered by the winter frosts and winds. It should be friable enough for seed sowing to commence towards the end of March.

Making the border

During winter a plan should be made, and the same remarks also apertain to where annuals are being sown in the herbaceous border, by which the plants will be grouped in degrees of varying heights, beginning at the back with groups of sweet peas, the tithonia, the taller nicotianas, the annual lupin and sunflower, each growing to a height of between 4-6 ft. then come the taller antirrhinums, larkspur and cosmeas, groups of the zinnia, salpiglossis and sweet scabious, all growing to a height of just under 4 ft. Further to the front come the African Marigolds, lavatera, annual chrysanthemums and the various asters. Then the calendulas, stocks, venidium, nigella and the new dwarf nicotianas; with the eschscholtzias, dwarf cornflowers, calendulas and collinsia still more to the front. Then come the French marigolds, linaria, nemesia and Virginia stock, mignonette and the ursinia, the more compact antirrhinums and zinnias. Finally for an edging, the Tom Thumb antirrhinums, Waldersee aster, the mesembryanthemum, limnanthes, alyssum and lobelia. If the spring flowering biennials are incorporated, then these too should be arranged in similar order and will commence the display at the beginning of April, those hardy annuals being sown about that time without any protection, obtaining valuable shelter from the biennials.

Some thought should be given to flowering periods so that plants are raised which will provide a display of colour over as long a period as possible, and colours should also be considered with care. A delightful border may be made up in shades of blue, pink, and crimson, omitting the more brilliant and harsher reds and yellows.

Or the plants may be sown in such a way as to give strips of various colours, with yellow flowering plants at a certain height, orange at another, red another. There are numerous colour schemes, but whichever is chosen always plant in groups to obtain the best effect. Odd plants dotted about will never produce much of a display. It is better to make the choice of plant small annd use large groups rather than to make use of one or two plants of a large number of varieties.

A border may be made entirely of antirrhinums beginning at the back with those which attain a height of 3 ft and using the Tom Thumb strain for an edging. Here again strips of one colour may be used to great effect.

When sowing the rule should be to sow thinly and only just cover the seed which in many cases will be very small. Thick sowing is not only wasteful but with many annuals which resent transplanting, to sow thickly will mean that the plants cannot be used elsewhere with any degree of success.

To help with thin sowing, the seed of all annuals having very small seed should be mixed with a small amount of fine sand before sowing. If shaken up in a tin, the friction will cause some slight chipping of the skin and enable germination to be more rapid. Sow on a calm day so that the seed will not blow away, then carefully rake it into the soil which will be friable before the seed is sown.

Watering should not be necessary until perhaps in early May when dry conditions often prevail and the young plants must be kept growing. Watering should then be done during late afternoon to allow the moisture time to dry off the seedlings before night and the possiblity of frosts.

A very worth while idea is to place small twiggy sticks, about 12 ins long, about the place where the seed has been sown, pressing them into the soil this will not only afford valuable protection from cold winds but will ensure that those plants like Virginia stock and Phlox drummondii, which possess a less sturdy habit than others, are given support so that the bloom may be seen to the greatest advantage. though at first the twigs may appear unsightly the plants will soon grow about them completely hiding them by the early summer. Even where barn cloches are being used the twigs should still be inserted.

A bed or border of mixed annuals should be a mass of bloom, with little soil showing, but any overcrowding should be guarded against or the plants may, in a wet summer, tend to damp off. Overcrowding will also produce a weak, loose-jointed plant, one which will have made a preponderance of stem rather then bloom and foliage, whilst staking may be necessary to prevent the plants from falling over with the first strong wind. Even where the seed has been sown thinly it may still be necessary to thin out those plants which make bushy growth, like calendulas and African marigolds. Much will depend upon the habit of the plant and upon whether it will transplant readily. Seed of the more vigorous of those plants which do not like moving, should be sown even more thinly than for those which make but little growth, generally those plants used for edging or for the front of a border.

Chapter 9

ANNUALS WHICH ARE SOWN WHERE THEY ARE TO BLOOM

ALL annuals will bloom from an outdoor sowing though for some this is the only satisfactory method, for they resent root disturbance. They are therefore grown by sowing the seed directly in the open as soon as the weather and condition of the soil permits. In this chapter those annuals which are generally raised under glass or from an early autumn sowing and are later transplanted will not be described here, as they appear elsewhere in this book whilst those which are generally grown for cut flower purposes, e.g., nigella, annual chrysanthemums and larkspur, and which are always happier sown where they are to bloom were described in Chapter 7. Thereare many plants which are grown in beds or about the herbaceous or annual border or shrubbery where they should be given the culture as described in the previous chapter. They are amongst the most inexpensive of all garden plants, a glorious summer display being obtained from only a few packets of seed.

f. Denotes a Fragrant Plant.

ADONIS (h). With their anemone-like flowers and their serrated foliage, this plant should be more widely grown.

There is both a summer and an autumn flowering species, but as the former, *A. aestivalis,* is slow to germinate, seed should be sown in late August the previous year and this may mean that it should be confined to a southern garden. Seed of *A. autumnalis,* is sown early in April. Thin out the plants to 10 ins apart. They reach a height of 12 ins.

ALONSOA (hh). This lovely annual may be raised in gentle heat or sown late in April where it is to bloom. The dainty flower spikes resemble miniature sidalceas. The best form is *A. Warscewiczii compacta,* which grows to a height of 15 ins. and should be allowed the same distance to mature. Pinch back the growing point to make a bushy plant.

ANCHUSA (h). The annual form *A. capensis,* Blue Bird, makes a bushy plant 15 ins tall and covers itself with brilliant blue flowers resembling forget-me-nots. Sow early April for it is completely hardy.

BARTONIA (h). f. The large golden-yellow blooms of *B. aurea,* with their numerous fluffy stamens give this flower a most attractive appearance. This is a valuable plant for a dry, light soil and has a long flowering period, mid-June until late September. Sow early April and thin to 1 ft apart, for the plants attain a height of 18-20 ins.

BRACHYCOME (hh). Known as the Swan River Daisy, this native of Australia enjoys a light, dry soil and a sunny position. It attains a height of only 9 ins and bears its blue, purple and lavender blooms from mid-June until October. Sow early in May and thin to 6 ins *B. iberidifolia,* Blue Star hybrids, is the best strain.

CACALIA (h). The Tassel Flower, a native of Africa, should be given a position of full sun and a dry soil. The plant bears tassels of bright scarlet, *C. coccinea,* whilst there is an equally striking golden form *C. aurea.* Seed is sown late in April in the open or late March in gentle heat as this plant transplants readily. As it grows to a height of 18 ins it should be thinned to 12 ins apart.

CANDYTUFT (h). The most easily grown of all annuals and much loved by children. The heads,like large pennies appear throughout summer in profusion, on 10 ins stems. It flourishes in any soil and may be used to the front of a shrubbery or almost anywhere. Obtainable in crimson, rose or lilac, the best strain is the Giant Hyacinth-flowered.

CLARKIA (h). Growing to a height of 2 ft. it is ideal for the centre of a border. The modern varieties have made this plant everybody's favourite, the flowers being fully double and borne over a long period. Amongst the best are Firebrand, coppery-red; Enchantress, salmon-pink; Vesuvius, orange; and Chieftain, pale mauve. The plants should be supported by long twiggy sticks. No thinning is necessary if sown thinly.

CLARY (h). Annual salvia, *C. Horminum,* Oxford Blue, a charming plant, growing to a height of 18 ins. and bearing masses of deep blue flowers from mid-July until late September. There is also a lovely shell-pink form, Pink Gem. Sow early April and thin out to 12 ins. apart.

COLLINSIA (h). *C. bicolor,* Salmon Beauty, flourishes anywhere, producing masses of dainty salmon-pink flowers from June till the end of August on 12 in. stems. If the seed is sown thinly in early April, or in early autumn in a sheltered garden, no thinning will be necessary.

CONVULVULUS (h). For quickly covering a fence or trellis at the back of a border *C. major,* Morning Glory, with its beautiful trumpets in shades of blue and mauve is most effective. For the front of a border *C. minor* in its various forms, making a compact plant 10 ins. high is equally striking. Sow early April and thin to 10 ins. apart.

CORNFLOWER (h). *Centaurea cyanus* is the annual cornflower, so easily raised from a sowing made either in autumn or early in spring, either where it is to flower or for transplanting. Growing to a height of 3 ft. the Double Rose and Blue flowered is sown at the back of a border, the dwarf flowered to the front.

DIMORPHOTHECA (h). Though a native of South Africa, the Star of the Veldt is almost completely hardy and may be sown mid-April where it is to bloom or a month earlier in gentle heat. The plants grow to a height of 12 ins. and bloom throughout late summer, *D. aurantiaca,* bearing flowers in shades of orange and yellow. This plant needs a sunny situation and a light, sandy soil.

ECHIUM (h). At their best in poor, dry soil *E. planiagineum,* Blue Bedder, making a compact plant 15 ins. high and in bloom from June to September, is a valuable plant and is so free flowering that it may be used for bedding. Sow early April, spacing the plants to 15 ins. apart.

ESCHSCHOLTZIA (h). Like the echium, this is a most colourful plant for a poor, dry soil. Sown early in April the seed germinates rapidly and the plants, growing to a height of about 10 ins. bear a profusion of flowers throughout summer. Amongst the best varieties are Dazzler, flame; Enchantress, rose and cream; Orange King; and Crimson Queen. Sown in groups about a shrubbery the plants will add rich colouring.

FELICHIA (h). *F. geriana,* with its sky-blue star-like flowers borne on 6 in. stems this is a plant deserving of greater popularity. Like all South African plants, it should be given a sunny position and a dry soil. In bloom from June until October it makes a most valuable and original edging plant.

GILLA (h). Bearing their heads of star-like flowers on 18 in. stems this is an excellent plant either for border display or for cutting. *G. capitata*, bears blue flowers. The seed may be sown either in early autumn or early in spring, the plants being thinned to 12 ins. apart.

GODETIA (h). Deservedly amongst the popular of all annuals which are sown where they are to bloom. There are tall flowering varieties like the orange, Kelvedon Glory, and the salmon-pink Sybil Sherwood, which attain a height of 2 ft.; intermediate varieties such as the crimson, Firelight and the snow-white, Purity, so lovely together; and the dainty dwarf Lavender Queen and Crimson Glow which make compact plants less than 12 ins. in height.

No annual will provide a more vivid display if given a sunny position and a soil containing a little humus. Sow early in April and thin to between 9-12 ins. apart, depending upon the height of the variety.

HELIANTHUS (h). *H. annus,* **the Sunflower, is a valuable annual for hiding an unsightly wall and may also be sown at the back of a border. There is a single and double form bearing rich golden blooms and attaining a height of 8 ft. in a single summer. Not quite so tall is the striking Sutton's Red, the yellow bloom having a wide mahogany circle round the centre. Seed is sown early in April the seedlings thinned to 2 ft. apart. They will commence to bloom early August.**

JACOBAEA (h). This hardy plant, often listed as *Senecio elegans,* **is now obtainable in a double form, in crimson, purple, white and rose. The seed is sown at the end of March and as the plants attain a height of 18 ins. they should be thinned to 12 ins. apart.**

Senecio arenarius, **is the Annual Cineraria, of similar height and which bears its sturdy flowers through summer.**

LARKSPUR. **See Chapter 7.**

LAVATERA (h). *L. trimestris* is the annual Mallow, the deep rose-pink, Loveliness, growing to a height of 3 ft. and forming a dense bush-like plant, being the best. Requiring a moist soil, the seed is sown early in April and thinned to 2 ft. apart. The Mallow is at its best during August and September, a delightful plant for the back of the border.

LAYIA (h). This native of California is at its best in a dry, sandy soil and should be given a position of full sun. Seed is sown early in April to the front of the border for the plants grow only 10-12 ins. high and bear most of their bloom during August and September. *L. elegans,* Tidy Tips, the dainty yellow flowers being edged with white, is a most attractive plant.

LEPTOSIPHON (h). For edging, this is a lovely little plant, coming quickly into bloom from an early April sowing. The new French hybrids, reaching a height of only 4-5 ins., cover themselves in star-like flowers embracing all the colours of the rainbow. Coming into bloom early in June, they remain colourful until mid-September. Do no thinning.

LEPTOSYNE (h). Valuable for both cutting and garden decoration, with its semi-double marguerite-like flowers of golden yellow, Golden Rosette, growing to a height of 18 ins. The bloom is held on long, wiry stems and borne from mid-June until September from a late March sowing. Thin to 10 ins. apart.

LINARIA (h). The flowers of *L. Maroccana,* Fairy Bouquet, are like miniature antirrhinums, produced in profusion on 7 in. stems throughout mid-summer. The colour range is enormous, white, ruby, rose, purple, yellow, the spikes being neat and compact. Sow in any soil towards the end of March, no thinning being necessary if the seed is sown thinly.

LINUM (h). Growing to a height of

from 12-15 ins. and bearing dainty flowers of the most brilliant colours, continuity may be obtained if the first sowing is made towards the end of March, with a second sowing early in May. Thin to 6 ins. apart. *L. grandiflorum coeruleum,* bears bloom of brightest blue; *rubrum,* vivid crimson, and borne in great profusion.

LUPIN (h). Extremely hardy and growing strongly in the poorest of soils, *L. Hartwegii,* grows to a height of 2½ ft. and may be obtained in white, deep blue and rose. Seed is sown at the end of March and is sufficiently large to be planted individually about 15 ins. apart. *L. nana,* is the Tom Thumb lupin, growing to a height of only 15 ins. and useful for the front of a border or shrubbery.

MALOPE (h). Very similar in form and habit to the lavatera, *M. trifida* makes a bushy plant 3 ft. tall. The bloom is of various shades of crimson and rose, at its best from August until late October.

Sow at the back of the border early in April and thin to 20 ins. apart.

MARIGOLD (French and African). See Chapter 6.

MATTHIOLA (h) f. This is the Night-Scented Stock, so much appreciated if sown beneath a window which may be left open during a summer evening. *M. bicornis,* grows to a height of 10 ins. covering itself in tiny flowers of palest lilac. Sow early April, no thinning being necessary if the seed is sown thinly.

MIGNONETTE (h) f. This is one of those annuals which is especially valuable for sowing in a limy soil, it is in fact as great a lover of lime as the dianthus. It also likes a soil containing some humus. Though sweetly scented, the mignonette, *Reseda odorata,* has never, until recently, been conspicuous for the quality of its flowers.

NASTURTIUM (h) f. Most colourful in a dry, sandy soil where it does not make excessive foliage, no hardy annual provides more colour nor over so large an area in such a quick time. The Tom Thumb varieties growing to a height of 10 ins., may be used for bedding, spacing the large seeds 9 ins. apart and sowing early in April. Outstanding is Ryburgh Perfection, with its vivid scarlet flowers and silver foliage.

Almost as compact are the Dwarf Double Hybrids, comprising lemon, salmon, rose, orange and crimson shades and a host of intermediate colours.

For a terrace or for covering a bank, the Gleam Hybrids are most striking, being semi-double and sweetly perfumed. Outstanding is Fiery Festival, with blooms of luminous scarlet. All come into bloom mid-June and continue until early November.

NEMESIA (h). See Chapter 6.

NEMOPHILA (h). Enjoying a cool, moist soil and a position of partial shade, there is a semi-climbing species *N. aurita,* bearing deep mauve flowers; and *N. Menziesii,* Baby Blue Eyes, which bears pale blue flowers with a striking white centre. Growing to a height of only 6 ins. and coming into bloom mid-May from an early April sowing, this is an attractive edging plant.

NIGELLA. See Chapter 7.

OENOTHERA (h) f. It is *O. bistoria* which is the annual. It grows to a height of 20 ins. and bears large golden-yellow flowers, similar to those of the biennial form, the Evening Primrose. Seed is sown early in April, the plants coming into bloom late in July and remaining colourful until the end of September. Thin to 15 ins. apart.

OMPHALODES (h). *O. linifolia* is an attractive plant for the front of a border, its tiny cream flowers and attractive grey foliage providing a striking contrast to the more richly coloured subjects. Sow at the end of March and thin to 6 ins.

PAPAVER (h). It is *P. Rhoeas,* the Shirley Poppy, which is so valuable sown where it is to bloom, for the new double flowered strain embracing shades of plum, slate-blue, crimson, scarlet and many others add a richness to a border or shrubbery equalled by few plants. Seed is sown early April to bloom from July until late October, and a sowing may also be made in a sheltered border in early September to bloom mid-May until mid-August.

PHACELIA (h). Producing its pinnacles of Hair-bell-like flowers of a rich mid-blue, on 9 in. stems and coming into bloom early in June from an early April sowing, *P. Campanularia* is one of the most colourful of all edging plants for a dry, sandy soil. Thin to 4 ins.

PHLOX DRUMMONDII. See Chapter 6.

RUDBECKIA (h). Growing to a height of almost 3 ft., the annual form of the better known perennial, *R. bicolor,* is an excellent plant for the back of the border and also for cutting. The two best varieties are My Joy, large golden-yellow; and Kelvedon Star, the yellow blooms having a striking bronze disc. Seed is sown towards the end of March, the young plants being thinned to 15 ins. apart.

SALPIGLOSSIS. See Chapter 6.

SAPONARIA. See Chapter 7.

SPECULARIA (h). Venus's Looking Glass, as *S. Speculum* is called, is a delightful edging plant, from an early April sowing flowering from mid-July until September on 9 in. stems. The deep violet bell-shaped flowers sow themselves as readily as forget-me-nots and quickly come into bloom.

SWEET SULTAN. See Chapter 7.

TROPAEOLUM (h). There are two annual species of the climbing nasturtium *T. Lobbanum,* obtainable in crimson, orange, scarlet and gold; and *T. canariense,* the golden Canary Creeper, both of which will cover a trellis in a single season and which are happy in any soil. Sow at the end of March, 15 ins. apart.

URSINIA (hh). With their orange daisy-like flowers and the bushy habit of the plants the ursinias lend distinction to any border. They may be sown at the end of April where they are to bloom, or raised in heat for they will readily transplant. *U. anethoides* and its hybrids in shades of ruby and purple, bloom on 12 in. stems, with the orange flowered, Golden Bedder rather more dwarf. Allow 10 ins. between the plants.

VIRGINIA STOCK (h). Growing to a height of 5-6 ins. and obtainable in shades of crimson, mauve, white and blue, these dainty little plants are delightful for edging a border, sown at the end of March. No thinning is necessary if sown thinly.

VISCARIA (h). Any garden soil suits this plant but they do like an open situation. *V. oculata* and its hybrids reach a height of 15 ins. and may be obtained in shades of blue, mauve and deep pink, whilst the Tom Thumb hybrids in white, rose and blue, flower on only 6 in. stems. Seed is sown early in May and the plants thinned to 6 ins. apart.

Chapter 10

ANNUALS INDOORS

The Value of annuals in home and greenhouse — Cultural methods — Summer culture — Care of indoor plants — Raising the plants — Hardy and half-hardy annuals for pot culture.

The sturdy, compact habit of the modern annual plant makes it very suitable for use as a pot plant either in the home or the greenhouse. Not only will the use of these plants in this way prolong the flowering season of the annuals, but they provide an inexpensive way of beautifying the house, compared with the cultivation of cyclamen, begonias and other of the more expensive plants. And as with almost all annuals, they readily adapt themselves to indoor culture, presenting few troubles in their management.

There are various ways in which these plants may be raised, but all should be grown in pots. Certain annuals such as the sweet pea and large-flowered antirrhinum may be grown in a warm greenhouse to provide the market with early bloom, and in this way are planted in the border or in ground beds. They are a specialist's crop and do not come within the scope of this book, but for all other purposes annuals are grown in pots.

Certain plants which have a long flowering season and which are frequently still in bloom when the beds are cleared in late autumn, may be removed with a ball of soil and if potted and placed in a warm room, will continue to bloom possibly until Christmas. Two plants which respond to this method are the dwarf French marigold and the new button asters, Remo and Waldersee which will bloom outdoors until November 1st and almost until the year end indoors, providing colour at a time when cut flowers are expensive.

Other plants may be sown early in the New Year and brought on entirely as pot plants in a warm greenhouse or room and will come into bloom late in spring when the indoor bulbs have lost some of their sparkle.

There are others, such as the schizanthus and celosia, which though occasionally used for bedding are almost always grown as indoor pot plants to flower late summer from a spring sowing.

The same remarks also apply to those plants grown indoors as when grown in the open, some do not like transplanting and should be sown in the pots where they are to bloom or first into 2½in. pots and transplanting to a larger size with the ball of soil intact and the roots undisturbed. And there are others which may readily be transplanted and which are generally sown in boxes or pans and transplanted when large enough to handle.

It must be said that excessive heat for almost all annuals is not required, indeed a temperature above 52°F. (11°C) will be detrimental to the growing of a satisfactory pot plant and so long as the plants are given a temperature of not less than 40°F. (5°C), they may be brought to perfection of bloom. They will be happiest in a temperature of 45°F. (7°C), but will grow sturdy and compact in temperatures between 40°-50°F. (5°-10°C).

Cultural Methods

Before making any sowings, it is advisable to consider the hardiness of the plants and the conditions which can be provided. Where no heat is available and the greenhouse has to depend upon the sun's rays in early spring for warmth, then it is advisable to grow only those plants which may be sown in early September and are generally wintered under cloches or in a cold frame. Or they may be sown and kept in the cold greenhouse, keeping the plants almost completely dry over the winter and moving to pots mid-March. They will come into bloom late in May, but would flower six weeks earlier if gentle heat, say 42°F. (5°C) could be provided. Only those plants should be grown in this way that are sufficiently hardy to withstand what may be a severe winter, e.g. calendulas, cornflowers, candytuft.

Owing to the smaller area of glass and the correspondingly reduced amount of air space, a cold frame is better than a cold greenhouse in which to winter these hardy annuals. It may also be given more substantial protection by covering the lights with mats or sacking, even old carpeting; and by heaping soil around the frames, possibly with the addition of hot-bed compost during January.

Nothing like so effective measures can be given those plants growing in a cold greenhouse, but the plants may be covered with brown paper in January, removing it whenever possible, whilst lengths of hessian canvas may be fastened to the inside of the roof.

Where growing on a large scale, possibly for sale, plants being raised in a greenhouse may be protected by using one of the new fumeless, hot-water radiator oil stoves. They are most inexpensive and need be used only for about ten weeks, from Christmas until early March, when the weather is most severe. Provided it is possible to maintain a temperature of 40°F. (5°C) during this period, the less hardy plants may also be raised for early bedding, or for indoor flowering in pots, but all those colourful annuals, such as Tom Thumb antirrhinums, dwarf cornflowers, etc., may be brought into bloom early in April, or at least ten weeks earlier than those plants in the open ground.

Where no greenhouse is available, then the more easily grown plants, i.e., those which germinate quickly and do not require the maximum amount of light, may be raised by sowing seed under cloches or in a box covered with glass at the end of August, or in pots under a barn cloche, and the young plants being wintered in a light, airy porch or room. Seed may also be sown in the same way early in April, of plants such as the celosia and schizanthus, which will come into bloom in mid-summer and continue until the marigolds and dwarf asters are brought indoors at the end of autumn, thus providing eight months of indoor bloom.

More use may also be made of various biennials, such as forget-me-nots and the extremely hardy winter-flowering pansy. They are lovely subjects for pots, lifted from the open mid-October and placed in the sunny window of a cool room; like the annuals they must be grown quite cool.

Summer Culture

All annuals may be flowered in pots, that is those with neat, sturdy habit, from an early spring sowing made as soon as there is some warmth in the sun and many will come quickly into bloom from seed. But it must be remembered that the same remarks apply for spring and summer culture as where the plants may be growing in a warm place during winter. It is important to provide as cool a condition as possible, even for the most tender varieties. All annuals are by nature hardy plants and will become drawn and loose jointed and never give a good account of themselves if grown in excessive heat. For this reason these hardier plants should be sown in autumn if possible, so that they will bloom indoors before the weather becomes really warm, possibly early in June. Then for the mid-summer period, those plants which are able to withstand more heat are brought into flower. If the more hardy annuals cannot be brought into bloom very early in summer, so that they will have finished by about July 1st, then it is better not to grow them and to concentrate on the schizanthus, celosia and similar plants which will bloom well under glass through summer.

Another method for growing annuals in pots is to sow seed of the half-hardy varieties in mid-summer either in a cold frame or in a greenhouse, or where neither is available, in boxes under a sheet of glass. If sown about the 1st of July, and grown on in pots the plants will commence to bloom early September and will continue until the year end.

At all times the plants must be kept as cool as possible and never allowed to suffer from lack of water. It must be remembered that plants in pots will dry out much more quickly than will plants in the open ground.

After transplanting, the plants should be placed in a partially shaded position, where the mid-day sun is not too severe.

Care of Indoor Plants

The summer care of annuals in pots should consist of attention to watering and ventilation. A warm, stuffy atmosphere will be completely against the raising of a satisfactory plant, so whenever conditions permit, ventilate freely, during the winter keeping closed those ventilators on the side of the prevailing wind, for draughts are harmful to young plants. Nor is it advisable to admit fresh air whilst the weather is damp and foggy. During the mid-winter months the plants should be kept as dry as possible. If their foliage is given a dusting with green flowers of sulphur in November and again in January, this will prevent damping off. But water only when absolutely necessary during the November-March period, possibly not at all, stepping up watering when the sun becomes warmer.

During the summer months the plants may need watering twice each day and will appreciate the spraying of their foliage when the weather is hot. When watering the pots, give a thorough soaking so that the roots do not have to turn upwards to the surface in search of moisture. In winter, pot plants will require watering possibly only once each week.

As an additional means of protection from the sun, the glass of conservatory and greenhouse should be whitened with lime to divert as much of the sun's rays as possible.

Raising The Plants

The method of culture is quite simple. Upon how the plant behaves when transplanted will depend upon how it is raised. Those plants easily transplanted will generally be sown in boxes or pans, sowing thinly for over-crowding often brings disaster. Those not easily moved must either be sown in the 5 ins. size pots where they are to bloom or in 2½ in. pots so that when necessary they may be moved with the soil intact.

When sowing in pots several seeds are sown to each, two to a 2½ in. pot, four to the 4 in. size, and six to the 5 in. size, but here again much depends upon the habit of the plant. With those plants making bushy growth less seed is sown or they should be so thinned that only one or two seedlings remain to each pot. Transplanting or thinning must be done as soon as possible for at all times it is essential to build up a sturdy plant.

Where the seed is sown in boxes or small pots for re-potting later, the John Innes Seed Sowing Compost should be made up, using sterilised soil where possible; and for sowing in pots where the plants are to bloom, or as a re-potting compost, use soil-less compost or John Innes Potting compost made up of:

7 Parts Loam
3 Parts Peat
2 Parts Coarse Sand
4 ozs. J.I. Base
¾ oz. Chalk
Per Bushel

John Innes Base consists of:

2 Parts hoof or horn meal
2 Parts Superphosphate
1 Part Sulphate of Potash
By weight

These quantities have been arrived at as the result of detailed experiments and should never be exceeded.

Never fill the pots right to the brim or watering will prove difficult and inadequate. Always use clean pots and see that before filling with compost they have been given some crocks at the bottom to ensure drainage. The compost should be kept dry until used, it being watered only after the seed has been sown. The plants are grown on as described, no hardening being necessary unless they are to be planted out. As mentioned in an earlier chapter, those more tender plants which do not like root disturbance are best raised in small pots in gentle heat, being planted out early in June after hardening in cold frames. The zinnia is one plant which responds to this culture.

As it is necessary to build up a bushy plant when growing in pots, the main growing points of most should be pinched back when the plants are about 4 ins. high. This will enable them to produce new shoots from the base. There has of recent years been a great improvement in the habit of most annual plants but even so a number possess a more delicate, struggling habit and they will require a number of small thin twigs or pot sticks for support. They should be given when the plants are about 6 ins. tall and before they begin to fall over. Those that grow fairly tall may require staking and tying in the orthodox way. The supports, which should be neat and clean, will be soon lost in the foliage and bloom.

Though pot grown annuals should never be given more than a small amount of nitrogenous manure, the J.I. Base contains only 5 per cent, pot grown annuals do respond to an occasional application of dilute manure water when in bud. This will greatly enhance the colour and quality of the bloom. And as with annuals used for bedding the dead blooms should be carefully removed as they begin to fade.

Hardy Annuals Suitable For Indoor Pot Culture

ANTIRRHINUM. Really a perennial but like each of the hardy annuals used for early flowering indoors, it is given biennial treatment, sowing the seed towards the end of summer and growing on in a frame or cold greenhouse. For pot culture, the semi-dwarf or intermediate varieties, growing to a height of 16 ins. are most suitable. The plants should be grown one to a 60 size pot, the leader shoot being pinched out to encourage a bushy plant. The Tom Thumb varieties are also very charming grown in pots.

CACALIA. *C. occinea,* with its vivid orange tassel-like flowers borne on 15 in. stems is a showy plant for a greenhouse or for the home. It does best made to flower in autumn from an early summer sowing.

CALENDULA. From an early autumn sowing under cloches or a frame, the plants may be lifted and potted mid-March when they will come into bloom early May. For pot culture sow the compact Radio, bearing orange flowers with quilled petals.

CANDYTUFT. A few seeds sown in a 48 pot early in March will come into bloom early in May and prove colourful for several weeks.

COLLINSIA. This showy plant may be raised from seed sown early autumn for late spring flowering it being one of the earliest of all annuals to bloom; or from a June sowing the plants will bloom late in summer. For pots the best is *C. bicolor,* Salmon Beauty, which grows to a height of 12 ins.

CORNFLOWER. It is the dwarf, double varieties which are so valuable for pot culture. The seed may be sown in early September, or in gentle heat in February. The plants are potted early April and will bloom in May, or outdoor plants may be lifted when still in bloom. Like the dwarf marigolds and asters they will continue to bloom indoors.

GODETIA. Sow September or early March keeping the plants as cool as possible. Two seeds should be sown in a 60 size pot selecting the best plants for growing on. Select the dwarf Crimson Glow and Lavender Queen for pot culture for they do not exceed 12 ins. and make stocky plants.

LEPTOSIPHON. Coming into bloom early in May from a mid-March sowing, the dainty French hybrids, flowering on 6 in. stems, and embracing a wide colour range are colourful plants for early in summer. Sow several seeds around a 48 pot.

MIGNONETTE. Sow the seed in small pots in early September wintering in a cold frame and transferring to greenhouse or home mid-March for blooming early May. The fragrance is particularly appreciated in the home. The most compact variety for pot culture is Red Monarch.

Half-Hardy Annuals Suitable For Pot Culture

AGERATUM. Raised in gentle heat in February and transplanted to small pots of 3 in. size, the bedding varieties are very pleasing to bloom through early summer. *A. mexicanum,* Purple Perfection, growing to a height of 10-12 ins. makes a delightful autumn flowering plant if seed is sown mid-summer and the plants grown on in 60 size pots.

ALONSOA. The species *Warcewiczii compacta,* with its bright scarlet blooms, produced on 18 in. stems is a valuable pot plant for autumn flowering from a mid-summer sowing.

ASTER. It is the Lilliput or Button-flowered varieties growing to a height of about 15 ins. which are so useful for pots for late summer and autumn flowering. The plants may be potted from the open ground and taken indoors mid-September, or seed may be sown mid-June and the plants grown on in pots. The new Remo and Waldersee strains are especially valuable for pots.

BALSAM. More correctly the Impatiens, this is a delightful summer-flowering plant, which may be grown on indoors from a mid-March sowing. Two forms are useful for pots, in fact no lovelier pot plant exists. *Impatiens*

Balsamina, in the rose and camellia-flowered strains; and *I. Holstii,* the hybrids obtainable in shades of rose and orange. Both grow to a height of 18 ins. and though of sturdy habit require staking.

BROWALLIA. It is *B. elata,* with its attractive glossy foliage and deep violet flowers, growing to a height of 18 ins. which is so valuable for indoor culture. From a late March sowing the plants bloom late in summer and continue through autumn. Transplants readily.

CALCEOLARIA. That much neglected plant *C. mexicans,* which from April sowing comes into bloom late in July and continues until October, bearing its pale yellow slipper-like flowers on 12 in. stems. The seed is minute in size so do not cover when sown, except to cover the box with a piece of glass.

CELOSIA. This is one of those tender annuals which prefer greenhouse treatment. Sow the seed in heat mid-March or in a cool house early April. The plants like a temperature of 65°F. and a humid atmosphere and may not easily be grown with other annuals requiring much cooler treatment.

CLINTONIA. *C. elegans,* with its deep blue flowers like a miniature flax, having an attractive white mark on the lower petal, is a charming little plant. Several seeds should be sown in a 60 size pot late in March when the plants will bloom from mid-June until early autumn.

DIASCIA. Like all South African plants, *Diascia Barberae,* Salmon Queen, with its dainty flowers almost like short bells, in bloom during mid-summer is ideal for bedding if given a dry soil and a sunny position. With its compact habit it also makes a charming greenhouse plant. The seed is sown early in March, the plants being grown on in small individual pots.

IPOMAEA. *I. rubro caerulea,* which

rapidly attains a height of 6-7 ft. and should be sown in gentle heat early in April and trained up the roof of a greenhouse or conservatory where its lovely sky-blue flowers borne throughout late summer and autumn are much appreciated. This plant is the convolvulus; a refined version.

MARIGOLD (African and French). These are amongst the most valuable of all annuals, almost hardy, but not quite, and so seed is sown under glass early in March for planting out at the end of May, or to be used for summer flowering pot plants. But they are more appreciated through autumn, when they will be in bloom from a late May sowing in a greenhouse; or plants may be lifted from beds in October with a ball of soil and they will continue to bloom for several more weeks.

PETUNIA. Ideal pot plants if only they did not attract Aphis so readily. The plants like a humid atmosphere; so do Aphis. From the beginning the plants should be regularly dusted with derris powder. The seed is sown in heat in February, the seed being small, and it is better not covered. Just press it into the compost and cover with a sheet of glass. When large enough to handle the seed is pricked into 2½ in. pots and again into the 60 size in which they bloom from July until October. The plants respond to a small amount of decayed manure in the final potting compost. For potting the brightly coloured Rose of Heaven; the terra-cotta Fire Chief; and Silvery Lilac are compact and free flowering.

SALPIGLOSSIS. Though rather taller growing than most other annuals used as pot plants this is such an outstanding plant indoors that it is worth while staking. Seed is sown in gentle heat in early April, the plants being grown on in 48 size pots to bloom throughout late summer.

SCHIZANTHUS. The Butterfly Plant or Poor Man's Orchid, is easily grown, but to grow those compact, bushy plants attention to detail must be given. Sow in gentle heat early March, pricking off the seedlings to 2½in. pots as early as possible. As the plant makes growth stop once or even twice, and stake to prevent untidiness. Keep the plant growing and water with dilute manure when the buds form. A most striking variety is Crimson Cardinal, growing to a height of 16 ins., whilst Dwarf Bouquet, more dwarf is obtainable in a wide range of colours.

STOCK. Extremely valuable for winter flowering under glass as pot plants, indeed, one may enjoy an almost all-the-year-round display from stocks, providing an almost continuous display from the New Year until the East Lothian Stocks have finished flowering late in autumn.

Both the Brompton and East Lothian stocks are biennials, the seed being sown under glass at the end of June, the Bromptons coming into bloom in the early New Year, the East Lothians in early spring.

A sowing of the annual 10-week stocks may be made in February to provide summer bloom, with a second sowing of the East Lothians being made late in March to come into bloom in September and to continue until the New Year, when the Brompton's come round again.

As stocks are liable to Black-Leg, it is either essential to use sterilised soil or to water with Cheshunt Compound. In any case always keep the plants as dry as possible through every stage. Sow thinly and transplant as soon as large enough to handle. No stopping or pinching is necessary for the plants have a naturally branching habit.

TORENIA. Almost like miniature antirrhinums *T. Fournieri,* its violet blooms having an attractive yellow throat, makes a lovely pot plant. It grows to a height of 12 ins. and from a mid-March sowing in heat, blooms from early July until October.

Chapter 11

ANNUALS IN THE ROCK GARDEN

The Value of annuals on a rockery—Sowing and planting—Plants suitable for a rockery.

Once again the inexpensiveness of annuals and the rapidity from which they may be brought into bloom gives their use considerable value about the rock garden. But of even greater value is the fact that where a rockery is planted almost entirely with annuals, the plants may be removed in their entirety at the end of autumn. The rockery may then be completely cleared of all weeds to make way for the winter and spring flowering bulbs. Though the use of perennial plants for the rockery is generally accepted as the most popular method of providing colour, the combination of bulbs and annuals, the bulbs being planted at sufficient depth so that they are not disturbed when the soil is cleared of weeds, is becoming encreasingly popular during the present time of shortage of labour in the garden. For there is no more tedious job about a garden than trying to disentangle weeds and rock plants which have made excessive growth, and without disturbing their natural habit. It is a slow tedious business which often requires attention two or three times a year. With annuals and bulbs the rockery may be made just as colourful all-the-year round and may be easily cleaned at a single operation during autumn to be replanted the following spring and early summer.

Though annuals on a rockery should be given a soil which is retentive of summer moisture, it will not be necessary to provide the plants with the soil conditions as provided for those plants which occupy the ground for possibly a decade or more. In fact there are numerous annuals suitable for the most difficult soils which will provide a brilliant display with the minimum of attention. But only those plants should be used which have a neat, compact habit. The rampant growers should be omitted for they will crowd out those with a more dainty habit and their charm about the rockery will be lost.

There is yet another point in the use of annuals about a rockery and that is when the plants are removed in October, the pests go with them and the next season's plants are rarely troubled by those pests which so often cause havoc to perennial plants, e.g. slugs and woodlice.

Sowing and Planting

To plant a rockery, seed may be sown about the crevices made by the stones or in the open spaces, but try to sow in such a way that when the plants have matured they will take on as natural an appearance as possible. And when setting out plants which may have been raised in heat or under lights in the usual half-hardy way, always plant in groups rather than as odd plants about the rockery. When planting in groups or drifts, plant self colours, so arranging them that there is a drift of pale blue next to one of shell-pink, then use a crimson with a mauve, yellow and orange together, colours which intermingle and yet provide generous splashes of colour.

Sowing and planting is done exactly as described in previous chapters bearing in mind the site and situation and prevailing weather conditions. Always plant when the soil is moist, yet friable and rather than lose the plants wait until the frosts and cold winds have disappeared.

Cloches may be found most valuable about a rockery using them to cover early sown seed and to give protection, by placing them on their side, for the more tender plants.

Construction of a rockery is given in detail by the Author in *Rock Gardening* in this series.

It is important that when the rockery has been made, the soil be allowed several weeks to consolidate before planting out the annuals. And always plant firmly, allowing the plants room to spread. Too close planting will cause the plants to become weak and drawn, they will choke each other as well as the weeds.

Throughout the summer they will require little attention other than occasional watering during a dry period and watering with dilute manure water during July and August will enhance the quality and colour of the bloom.

Plants Suitable For The Rockery

AGERATUM (hh). The little ageratum which requires half-hardy treatment, sowing in boxes of sandy loam in gentle heat, or in a frame for later flowering and planting out early June, are charming subjects. Little Blue Star grows only 4 ins. tall and makes a rounded, compact plant; likewise, the salmon-coloured Fairy Pink. Plant with them the pure white Little Dorrit.

ALYSSUM (h). Though quite hardy the Sweet Alyssum is generally given half-hardy treatment for the seed is slow to germinate. The best varieties for a rockery are *A. minimum,* pure white, and the purple, Royal Carpet, both of which bloom at a height of only 3 ins.

ANAGALLIS (h). The Rock Pimpernel, producing masses of rich gentian-blue flowers. Sown round a stone they will cover it with blooms throughout summer and autumn.

ANTIRRHINUM (h). The bedding varieties are quite out of place on the rockery, but the Rock Hybrids have been evolved from *A. glutinosum* and are charming subjects, growing only 6 ins. tall and bearing masses of tiny "snap-dragons." They are best sown in boxes first, in gentle heat in February or they may be sown late in August and wintered in a cold frame for planting out during May.

CALANDRINIA (hh). Because its tiny crimson flowers open only when in the direct rays of the sun, this dainty plant must only be grown in a position of full sun. *C. umbellata* is a charming plant on the rockery growing to a height of 6 ins. and the same distances across. Sow the seed late in April where it is to bloom and thin to 6 ins. apart.

CHEIRANTHUS (hb). The mauve alpine wallflower of compact habit. *C. linifolius,* is really a hardy biennial and should be sown in July to bloom the following spring. It is a good plant to mix with small bulbs.

CONVOLVULUS (h). *C. minor,* Lavender Rosette is the one to grow. It is of prostrate habit, but does not climb or trail about the rockery. Its grey foliage and pale lavender flowers make it an attractive subject which should be sown in early April.

COTULA (hh). The dainty *C. barbata,* growing to a height of only 6 ins. and bearing its tiny lemon-yellow blooms throughout late summer and autumn is most attractive used with drifts of alyssum Royal Carpet. It is half-hardy but transplants well.

ESCHSCHOLTZIA (h). The variety, Miniature Primrose, growing no taller than 4 ins. and bearing dainty primrose-yellow flowers right through summer is an ideal plant for a rockery or for edging. Sow where it is to bloom in early April.

GAMOLEPSIS (hh). In its habit and bloom almost identical to the tagetes, bearing tiny golden flowers over the same long period and at a height of 8 ins. Treat in the same way as for the ordinary tagetes.

GRAMMANTHES (hh). Very similar in form and habit to the mesembryanthemums, *G. gentianoides* has almost a prostrate habit and bears flowers of scarlet, orange and yellow and intermediate shades. Like all South African plants it likes a dry soil and a position of full sun. Sow in heat mid-March, planting out early in June. The plant is most colourful from the end of July until November.

IONOPSIDIUM (h). The Violet Cress, growing no taller than 2 ins. and covering itself in a mass of violet flowers through summer. Sow in the crevices where it is to bloom.

LEPTOSIPHON (h). Sown early in April, this dainty plant grows no taller than 6 ins., yet forms a large mat which it covers with flowers of various shades. It enjoys a dry, sunny position.

LIMNANTHES (h). *L. Douglassi* is a

delightful hardy annual bearing its yellow and white flowers in profusion. Two sowings should be made where conditions are favourable, one in August to give colour in spring and another in April to provide late season colour.

LINARIA (h). *L. moroccana,* Fairy Bouquet is the best species for the little brilliantly coloured antirrhinum-like flowers are carried on but 6 in. stems. A hardy annual, this is one of the most charming of all annual alpine plants.

Another species *L. Cymbalaria,* the Kenilworth Ivy, really a perennial, may also be raised from seed to flower the same year.

LINUM (h). The species *Grandiflorum rubrum,* with its rich crimson blooms borne in profusion throughout summer, is a useful addition to the rockery, especially attractive with a limestone rock for a background.

LUPIN (h). Sown where they are to flower for they will not transplant well, the Tom Thumb lupins, growing to a height of only 8-9 ins. are most attractive sown in small groups about a rockery.

MARIGOLD (hh). Almost all the dwarf French and African marigolds are very suitable for a rockery, but the Cupid marigolds are quite outstanding. They make compact bushy little plants growing only 5-6 ins. tall and almost 9 ins. across and bear an abundance of double chrysanthemum-like flowers 2 ins. across. Treat as half-hardy, sowing in heat or under lights in March. The plants continue to bloom until November, always remaining neat and tidy.

MATRICARIA (hh). The compact *M. eximea,* covers itself with masses of golden ball-like flowers and is perfectly at home on the rockery. The plants are raised by sowing in gentle heat in March and planting out early May.

MESEMBRYANTHEMUM (hh). These dwarf South African plants will continue to bloom in a frost free garden and in a dry, sandy soil until Christmas, bearing an abundance of small rayed blooms in the widest possible colour range. *M. criniflorum* or the Livingstone Daisy is a half-hardy plant which likes a dry, sandy soil and a position of full sun. Even planted between crazy paving stone it will grow well sending out its succulent claw-like foliage and covering itself in masses of the most brilliantly coloured daisies.

NEMOPHILA (h). *N. insignis,* known as Baby Blue Eyes is a colourful plant with trailing habit and bright blue flowers with white eyes. It should be sown every month from April to June between crazy paving stone, or about the rockery where it will provide a colourful display throughout summer.

PHACELIA (h) *çampanularia.* This is a hardy plant that could be more often used on the rockery to advantage, for it is of compact carpeting forming habit, the navy blue bell-shaped blooms being most attractive. It is known as the Californian Bluebell.

PHLOX DRUMMONDII (hh). This is the half-hardy phlox, the new dwarf hybrids reaching a height of only 6 ins. and remaining delightfully colourful right into autumn. The new variety Isabelina is of a lovely yellow shade; Meteor, salmon; and Blue Bird, sky-blue. Two or three planted together look very showy.

PORTULACA (h). Growing to a height of 6 ins. this member of the Purslane group *P. grandiflora* is an excellent carpeting plant bearing large flowers in the most vivid shades of rose, yellow and scarlet. Sow where it is to bloom, thinning to 6 ins.

SAPONARIA (h). To fill in those odd spaces about the rockery, sow the white and scarlet forms of *S. calabrica* which bears its dainty flowers on 7-8 in.

stems. It grows rapidly from seed and sowings should be made in early April and in all those empty spaces until the end of July. Thin to 5 in. apart.

SWEET WILLIAM (h). This is a very dwarf, compact variety of this charming old world plant that should be suitable for the rockery. It is called Indian Carpet and may be obtained in a mixture of all those lovely salmon, red and white shades familiar with their taller cousins. They should be sown in frames in July to flower the following year.